Social Movements against Wind Power in Canada and Germany

Taking a comparative case study approach between Canada and Germany, this book investigates the contrasting response of governments to anti-wind movements.

Environmental social movements have been critical players for encouraging the shift towards increased use of renewable energy. However, social movements mobilizing against the installation of wind turbines have now become a major obstacle to their increased deployment. Andrea Bues draws on a cross-Atlantic comparative analysis to investigate the different contexts of contentious energy policy. Focusing on two sub-national forerunner regions in installed wind power capacity – Brandenburg and Ontario – Bues draws on social movement theory to explore the concept of discursive energy space and propose explanations as to why governments respond differently to social movements. Overall, *Social Movements against Wind Power in Canada and Germany* offers a novel conceptualization of discursive-institutional contexts of contentious energy politics and helps better understand protest against renewable energy policy.

This book will be of great interest to students and scholars of renewable energy policy, sustainability and climate change politics, social movement studies and environmental sociology.

Andrea Bues is a research analyst at the German Advisory Council on the Environment and is based at the Potsdam-Institute for Climate Impact Research. She holds a PhD in political science from Freie Universität Berlin and has been working on land use conflicts and energy transitions.

Routledge Studies in Energy Policy

For further details please visit the series page on the Routledge website: www.routledge.com/books/series/RSIEP/

Social Movements against Wind Power in Canada and Germany

Energy Policy and Contention

Andrea Bues

Routledge
Taylor & Francis Group
LONDON AND NEW YORK

earthscan
from Routledge

First published 2020
by Routledge
2 Park Square, Milton Park, Abingdon, Oxon OX14 4RN

and by Routledge
605 Third Avenue, New York, NY 10017

First issued in paperback 2022

Routledge is an imprint of the Taylor & Francis Group, an informa business

Publisher's Note
The publisher has gone to great lengths to ensure the quality of this reprint but points out that some imperfections in the original copies may be apparent.

British Library Cataloguing-in-Publication Data
A catalogue record for this book is available from the British Library

Library of Congress Cataloging-in-Publication Data
A catalog record has been requested for this book

ISBN 13: 978-0-367-51051-0 (pbk)
ISBN 13: 978-0-367-43955-2 (hbk)
ISBN 13: 978-1-003-00670-1 (ebk)

DOI: 10.4324/9781003006701

Typeset in Times New Roman
by Wearset Ltd, Boldon, Tyne and Wear

Dedication – To C.P.

Contents

Illustrations

Figures

Tables

Acknowledgments

This book would not have been possible without the support of many individuals and organizations. First of all, I want to warmly acknowledge all my interviewees, who sacrificed their time to participate in this project. Without your openness and trust, this research could not have been realized.

My sincerest gratitude goes to Miranda Schreurs from the Bavarian School of Public Policy at the Technische Universität München for her constant guidance, invaluable suggestions and personal encouragement. Many thanks also go to Christian Lammert from the John F. Kennedy Institute, Freie Universität Berlin. I am grateful to the Leibniz-Institute for Research on Society and Space (IRS) for funding the research from 2014 to 2017. Particularly, I would like to thank the former group of research department 2, headed by Tim Moss and Ludger Gailing. The research benefitted from such an inspiring research environment. I would also like to thank Wolfgang Lucht from the Potsdam-Institute for Climate Impact Research (PIK) and the team at the German Advisory Council on the Environment (Sachverständigenrat für Umweltfragen) for their support in helping me finish this project.

The field research in Ontario would not have been possible without the generous support of the German Academic Exchange Service (DAAD) for a research grant from May to July 2015 in the framework of their program for short-term doctoral grants and the Foundation for Canadian Studies for a research grant from October to December 2015. I also want to acknowledge the DAAD grant for conference participation to IPSA Montreal in 2014.

I am indebted to the scholars in Ontario who hosted and supported me during my three research stays: Ulrich Best at the Canadian Centre for German and European Studies, York University, Toronto; Jose Etcheverry at the Faculty of Environmental Studies, York University, Toronto; and Doug Macdonald at the School of the Environment, University of Toronto. All three offered their enthusiastic support and practical advice, while crucially helping to facilitate contacts into politics, academia and the renewable energy scene in Ontario. I also wish to thank Christy Hempel from Guelph University for her friendship, sharing her personal insights and inviting me to co-organize a workshop with practitioners and researchers on wind turbine planning in the Huran/Grey/Bruce area in Ontario. My gratitude also goes to the Communities Around Renewable Energy

Project (COAREP) group at Western University, Ontario, including at the time Jamie Baxter, Emmanuel Songsore and Chad Walker. Upon finalizing the book, Chad Walker had moved over to Exeter University and provided valuable comments to my Ontario chapters. Additionally, I wish to thank Joan DeBardeleben from Carleton University for a helpful early meeting on Canadian renewable energy in Berlin.

My work benefitted considerably from various panelists and participants of conferences and research colloquia. I am very grateful to the graduate student group at the Environmental Policy Research Centre (FFU) of the Freie Universität Berlin for providing a supportive and helpful forum to exchange early ideas and discuss progress. I am also indebted to my writing group: Gloria Amoruso, Andrzej Ceglarz, Yi hyun Kang and Dongping Wang. I would like to thank the team of the Energiekonflikte project at PIK for sharing their experience and discussing research progress. I am also indebted to Eva Eichenauer and Katherina Grashof for their valuable comments on earlier drafts of chapters. I thank Christina Lee for editing an earlier draft of the book. I thank my editors at Routledge, Annabelle Harris and Matt Shobbrook for their support of this undertaking, as well as three anonymous reviewers for their thoughtful comments.

Last, but not least, I wish to express my sincerest gratitude to my friends, Johannes and all other family members for their patience and support. Without their help my work would have been much less enjoyable.

Acronyms and technical terms

10h	Setback of ten times the height of a turbine to other land uses
AfD	*Alternative für Deutschland*
BUND	*Bund für Umwelt- und Naturschutz Deutschland*
CAD	Canadian Dollar
CDU	*Christlich Demokratische Union*
EEG	*Erneuerbare-Energien-Gesetz* (Renewable Energy Sources Act)
ERT	Environmental Review Tribunal
FIT	Feed-in tariff
GDR	German Democratic Republic
GDP	Gross Domestic Product
GEA	Green Energy Act
GHG	Greenhouse gas
GTA	Greater Toronto Area
IPCC	Intergovernmental Panel on Climate Change
kW	Kilowatt
LTEP	Long Term Energy Plan
LRP	Large Renewable Program
MAWT	Mothers Against Wind Turbines
MPP	Member of Provincial (or State) Parliament
MW	Megawatt
MWE	*Ministerium für Wirtschaft und Energie*
NABU	*Naturschutzbund Deutschland*
NAFTA	North American Free Trade Agreement
NGO	Non-Governmental Organization
NIMBY	Not-In-My-Backyard
NRWC	Niagara Region Wind Corporation
OPA	Ontario Power Authority
OVG	*Oberverwaltungsgericht* (Higher Administrative Court)
PC	Progressive Conservatives
REA	Renewable Energy Approval
RPG	*Regionale Planungsstelle (*Regional planning authority)

Setback	Required separation distances between wind turbines and other land uses
SPD	*Sozialdemokratische Partei Deutschlands*
STGB	*Städte- und Gemeindebund*
Wind suitability area	*Windeignungsgebiet*: Regional planning category; specifies areas designated for wind turbine development
WCO	Wind Concerns Ontario
WLWAG	West Lincoln/Glanbrook Wind Action Group

1   Introduction

Renewable energy politics and protest

Environmental social movements have seen a major revival. Within just one year, climate activist Greta Thunberg's weekly school strike turned into a new climate movement. The renewed public salience of mitigating climate change has prompted national governments to intensify efforts to meet the temperature targets of the 2015 Paris Agreement. The agreement unites 197 countries behind the goal of reducing emissions, keeping global warming well below 2° C above pre-industrial levels and pursuing efforts to limit temperature increase to 1.5° C. This practically means that global greenhouse gas emissions must peak no later than 2020 and then sharply decrease until 2050 (Rockström et al. 2017). As fossil fuels are among the major drivers of anthropogenic climate change, one could assume that phasing out fossil fuels and phasing-in renewable energy is to become a top priority for governments to meet the Paris Agreement and respond to public concerns.

Wind energy is among the most promising renewable energy technologies and has been regarded as a suitable source to replace high-emission fossil fuel-based energy systems. In 2017, renewables made up approximately 25.5 percent of global electricity generation, out of which 18 percent accounted for wind energy (IRENA 2019). A number of good reasons for governments to promote wind energy exist, besides their positive role in decarbonizing energy systems. The cost of wind power has declined substantially since its introduction (Veers et al. 2019) and on-shore wind energy is cost-competitive, or nearly so, with natural gas and coal-fired power plants in many regions (IRENA 2017). Harnessing domestic renewable energy sources such as wind power has also been regarded as a possible way to diversify a nation's energy portfolio, decrease dependence on fossil fuel imports and increase energy security. There are also benefits on domestic revenue and employment. Studies show that per kilowatt hour, alternative energy including wind and solar requires more than double the workforce in comparison to fossil fuel technology (Valentine 2015, p. 21). A variety of other advantages exist, including benefits for human health, the environment and ecosystems, as well as technological innovation (IPCC 2014; Luderer et al. 2019; REN21 2017).

Notwithstanding these benefits of renewable energy and wind power, wind turbines often meet with local disenchantment. Raised concerns usually relate to aesthetics, noise, impact on wildlife (especially birds and bats), shadow flicker or electromagnetic interference, financial loss and community unrest (Ellis and Ferraro 2016; Krogh 2011; Leung and Yang 2012; Thygesen and Agarwal 2014; Valentine 2015). Opposition to wind turbines is not a new phenomenon and has long been subject to discussion in academic literature and beyond (e.g. Alt et al. 1998; Wolsink 1988). Among the earliest approaches to explain anti-wind protest is the "Not-In-My-Backyard" (NIMBY) concept. NIMBY contends that opponents may be in favor of wind energy in general, but reject them in close vicinity due to individualistic, ignorant and selfish behavior. Despite its appealing simplicity to explain resistance to wind turbines, the NIMBY concept has been refuted and is now widely recognized as overly generalizing and insufficient to dismiss protest against wind turbines (Burningham 2000; Devine-Wright 2009; Jobert et al. 2007; Wolsink 2007).

Organizing rallies, putting up protest signs, lobbying politicians: anti-wind protests have turned into forceful social movements in many jurisdictions. In some instances, protest has become powerful enough to prompt governments to curb their renewable energy programs, as recent examples illustrate. In late 2019, the German government tabled a draft federal law that would severely restrict possible turbine sites. Around the same time, the Norwegian government backed away from a national wind energy plan. The government of Canada's most populous province, Ontario, cancelled support for renewable energy projects right after gaining office in 2018. Technological feasibility, lower costs and pressure from the streets to act on climate change are thus not the key predictors to the adoption, continuity and implementation of renewable energy policies. Other social and political forces play a part as well. They become evident in the discursive and institutional context against which disputes over wind turbines unfold.

Prevailing discourses in a society are one important aspect. Wind turbines can be framed as a viable solution to the urgent need to decarbonize energy systems – or as industrial constructions inflicting an overly heavy burden on rural areas. Both proponents and opponents of wind turbines may draw on arguments that embrace such publicly acknowledged discourses to support their cause. This is because both aim at displaying their opinion in a better light and seek to avoid appearing purely self-interested or profit-seeking. While opponents rather relate to the need to preserve the environment and the innate value of the proposed site, proponents invoke their contribution to the global fight against climate change (Haggett and Futak-Campbell 2011). If generally positive discourses on renewable energy and energy system change exist in a jurisdiction, it may thus be more difficult for anti-wind movements to garner support for their cause.

The existing institutional structure also has a major impact on how conflicts over wind turbines unfold. For instance, if local inhabitants perceive the local decision-making process over wind turbines as unfair, they may draw on arguments revolving around procedural injustice (Gross 2007). They may also refer to environmental or distributional injustice, which address the unequal allocation of environmental burden and benefits (Cowell et al. 2011). Institutional schemes that are perceived as just have consistently been linked with higher levels of social acceptance, for instance community energy schemes (Devine-Wright 2005; Ottinger et al. 2014; Wolsink 2007; Zoellner et al. 2008). However, it depends on the details whether these kinds of projects are perceived as just, for instance whether a sense of trust and similar understandings of justice exist between the involved actors (Goedkoop and Devine-Wright 2016; MacArthur 2016; Simcock 2016). A community energy scheme may therefore not be a guarantee for an embracing attitude toward wind turbines, but there is a greater chance of alleviating public disenchantment. Nonetheless, institutional structures play a key role in disputes over wind turbines and are also shaped by prevailing discourses on wind energy. While wind turbine development has for example taken the shape of small community schemes in Denmark, turbine development in the United States or the United Kingdom has developed as large-scale industrial projects. This difference is also due to diverging discourses around who should develop wind turbines.

Given the importance of the discursive and institutional context of wind disputes, the question arises how they matter for the response that anti-wind movements receive from governments. Decarbonizing energy systems is of utmost importance to meet climate targets and renewable energy represents an important technology to replace greenhouse gas intensive energy generation. It is therefore important to understand under which conditions governments will end previously successful renewable energy policies in the face of anti-wind protest. Many social movements start at the local level and therefore need to "scale-up" their struggle to the policy-making level. In order to do so, the movement needs to garner support in terms of being recognized "by its antagonists as a valid spokesman for a legitimate set of interests" (Gamson 1975, p. 28). Which discursive and institutional circumstances thus favor or hamper the scaling-up of protest, and how does the discursive dimension interact with the prevailing institutional structure?

Current approaches in the literature have fallen short of explaining the impact of both discursive and institutional context in the study of social movement's impact. The classical literature on social movements has provided different explanations of government responses to social movements. Common explanations for different movement outcomes are the external institutional environment conceptualized as the political opportunity structure (Eisinger 1973; Kitschelt 1986; Meyer and Minkoff 2004; Vrablikova 2014) and the discursive environment in form of media discourses (e.g. Aydemir and Vliegenthart 2017; Koopmans and Olzak 2004; Koopmans and Statham 1999; Motta 2015). Notwithstanding the importance of these contributions, they are not enough to explain why governments respond differently to anti-wind protest. We do not only want to know how either the institutional or the discursive structure matters, but we are interested in both as well as in their interaction.

Analytical approach

Discursive institutionalism (Schmidt 2008, 2012, 2015) and the argumentative discourse analysis approach (Hajer 1993, 1995, 2006) can help. Discursive approaches have been used widely in the study of environmental politics (Dryzek 1997; Feindt and Oels 2005; Hajer and Versteeg 2005; Leipold et al. 2019; Szarka 2004). Discursive institutionalism aims at explaining "the dynamics of change (but also continuity) through ideas and discursive interactions" (Schmidt 2011, p. 60) and offers valuable explanations on how institutions and discourses interact. Following the argumentative discourse analysis approach, disputes over wind turbines can be regarded as a "struggle for discursive hegemony" (Hajer 1995, p. 59). This characterization emphasizes the discursive interaction between anti-wind movements and governments and highlights the role of discourses and framings. These two approaches offer valuable insights on the interaction between discourses and institutions and can be used as a basis for conceptualizing how the discursive-institutional context supports or hampers challengers such as anti-wind movements.

The book departs from the literature on social movements and combines it with elements of discursive-institutional analysis and notions of power to elaborate the concept of "discursive energy space". The discursive energy space is made up of prevailing discourses and framings of the energy sector (the meaning context) and the formal institutional system, which consists of the way in which sub-national energy policy is connected to federal energy policy and the local space of citizen participation. In combination, the meaning context and the formal institutional system form the discursive energy space. The main argument is that discursive energy spaces provide "discursive strength" to the anti-wind movement, which considerably influences government response. Discursive strength is defined as the degree to which an anti-wind movement is able to scale-up its message from localized conflicts to the policy-making level by being accepted by the government as representing legitimate interests. The elements of the discursive energy space have different power effects that contribute to differences in a movement's discursive strength. Those include the first, second and fourth face of power (Flinders and Buller 2006; Foucault 2000; Lukes 1974). Developing the concept of discursive energy space, the study focuses on the following three research questions: *How do discourses and frames on energy transitions impact the discursive strength of anti-wind movements? How does the existing formal institutional system affect the discursive strength of anti-wind movements? How do these two aspects interrelate and influence government response?* In combination, these three questions help us understand how context matters in the response that social movements receive from governments. The aim of the study is to explore disputes over wind turbines at the sub-national level of two frontrunner jurisdictions in energy system change, as other jurisdictions can possibly learn from their approaches and experience.

The study contributes to an emerging body of literature on the politics of energy transitions, investigating the discourses behind and the strategies of those advocating the implementation of wind energy as one important avenue for energy transitions, and the resistance strategies of those opposing them. The emphasis here is on the discursive interaction between a sub-national government and an anti-wind movement. In this way, the study also helps build a better understanding of the conflictual nature of energy transitions. Most of the studies on anti-wind opposition have narrowly focused on single case studies exploring the reasons behind protest, while few studies have involved cross-country comparisons of the outcomes of anti-wind opposition.

Geographic focus and case studies

The geographic focus of the book is Canada and Germany and the political focus is on the sub-national level. Due to their high energy sector emissions, especially the world regions of North America and Europe could substantially benefit from increasing their share in renewable power. In fact, wind power has represented the largest source of new electric capacity additions in many recent years in Canada and the United States (Rand and Hoen 2017). The European Union has also been a growing wind energy market with ambitious goals for further development (Lacal Arantegui and Jäger-Waldau 2018). The sub-national level of provinces and local communities has become more important in the past decade to formulate climate policies and programs (Jänicke et al. 2015). It plays a crucial role in efforts to combat climate change and increase the rate of renewable energy. As anti-wind protest usually becomes evident first at the local level, the sub-national level is directly targeted and is often the place possessing the most leverage for responding to this protest.

The two sub-national jurisdictions of Ontario (Canada) and Brandenburg (Germany) serve as examples of forerunners that have made great efforts in renewable energy development. Ontario comes in first place in Canada with regard to installed wind power capacity, while Brandenburg scores second place in Germany. Many other jurisdictions are encouraged to follow their example, decarbonize their electricity system and expand the rate of renewables. The book discusses and compares in detail these two sub-national forerunner jurisdictions. Brandenburg is a scenic, rather poor rural state surrounding Berlin. For years, the jurisdiction has benefitted from support schemes for renewable energy, so much so that it is difficult to find an area in Brandenburg where wind turbines are not visible. Brandenburg has won, on three consecutive occasions, the prize for the German federal state best promoting renewables (Agentur für Erneuerbare Energien 2012). In late 2019, the electric vehicle company Tesla has proposed building a large automobile plant in Brandenburg due to its forerunner status in generating renewable energy. The region is central to the ambitious renewable energy plans of the federal government. Germany plans at least 80 percent of renewable energy in electricity consumption by 2050. This means that many more wind parks and solar systems will be needed. Yet, opposition to the further expansion of renewable energy is growing. In Brandenburg, this has taken the form of about 100 local protest groups against wind turbines. Their umbrella organization initiated several public referendums, calling for a larger distance of wind turbines to residential buildings and a ban on the construction of wind turbines in forests.

Ontario is Canada's most important province in terms of economic capacity. It is also a forerunner in energy system change, having eliminated coal from its energy portfolio in 2015. This represented the single largest greenhouse gas reduction initiative in North America (Government of Canada 2016, p. 10). Following the German example of renewable energy policy support, the provincial government introduced the first large-scale feed-in tariff for renewables in North America in 2009 (Stokes 2013). This sparked the development of hundreds of wind energy projects in South-Western Ontario, but protest against them evolved rapidly and forcefully. About 100 local protest groups, united under the umbrella organization Wind Concerns Ontario, called upon the government to end support for wind turbines, arguing that wind turbines would be too costly and entail adverse health impacts. The anti-wind movement in Ontario closely worked together with the opposition party and made wind turbines an important topic in several provincial elections.

Despite this forceful social opposition against wind power in both jurisdictions, the sub-national governments reacted differently. While the Brandenburg state government was reluctant in adopting the movement's demands and continued their policy approach to wind energy, the Ontario provincial government first substantially altered and ultimately curbed its renewable energy policy. How can these differences be explained? Which discursive and institutional factors have played a role? Important lessons can be drawn from the comparison of Brandenburg and Ontario for the further development of renewable energy policies around the world. The objectives of the book are to show how federal states become forerunners in renewable energy deployment, how different contexts influence the chances of anti-wind protest groups to be successful and how and why governments respond differently to anti-wind protest.

Renewable energy support schemes

At least 176 countries had targets for renewable energy by the end of 2016 (IRENA et al. 2018, p. 22). Generally, the spectrum of renewable energy support policies ranges from political announcements and vision statements to legally binding renewable energy targets prescribed in detailed roadmaps and action plans. A variety of renewable energy support schemes exist (Lewis and Wiser 2007; Saidur et al. 2010) and two general types can be distinguished (Kwon 2015). A feed-in tariff is an example of a *price-driven* type of support scheme, and auction schemes are an example of a *quantity-driven* mechanism. A *feed-in tariff* (FIT) is a fixed payment per kWh of generated renewable energy – usually of a specified technology such as solar power or onshore wind – for a time period of about 15–20 years. This pre-set price varies according to the technology, size, source and sometimes also resource quality (Lauber and Schenner 2011). Usually fixed at the beginning of the contract period, the price may decrease over the course of time. Subject to certain conditions, everyone who owns a renewable energy facility is eligible to participate in a FIT scheme and grid operators are obliged to feed-in their generated electricity. *Auction schemes*, by contrast, are procurement mechanisms in which energy producers compete for a tariff in an auction by submitting bids specifying the level of remuneration for a specific volume of capacity or power the bidder proposes to realize. Some countries carry out single rounds, others set up a comprehensive schedule of auction rounds over several years inviting bids for pre-set volumes of capacity or power. While FITs have been applauded for providing low barriers of entry to new actors and technology due to their low administrative complexity (Verbruggen and Lauber 2012), auctions have become very popular lately, mainly because of the promise of reducing costs (del Río and Linares 2014). Many countries that have changed their support system from FIT system to auction system have done so in order to reduce costs. This also applies to Germany and Ontario.

Methodological approach

Cross-national comparisons can lead to "fresh, exciting insights and a deeper understanding of issues that are of central concern in different countries" (Hantrais 1995, p. 11). Given the similar wind technology, geographical characteristics, policy environment and characteristics of protest in Brandenburg and South-Western Ontario where most of the wind turbines are constructed, the research followed a most-similar comparative research design. Brandenburg and Ontario were chosen as critical and paradigmatic cases (Flyvbjerg 2011) for energy system transformation. They are critical because if the observed characteristics and findings apply to those two forerunners, then they potentially also apply to other sub-national governments employing similar positions on wind turbines. As the study looks at one sub-national forerunner in Europe and one in North America, the cases are also paradigmatic in that they highlight "more general characteristics" (Flyvbjerg 2011, p. 308) of these two geographic regions.

The research relies on extensive qualitative data that were collected in 2015 and 2016. The research approach followed an inductive qualitative research methodology. The main sources of data included 48 interviews and five focus group discussions, document analysis and participant observation. Data was generated at two levels. As the focus is on sub-national government responses to the demands of the anti-wind movements put forward by anti-wind umbrella organizations, the sub-national level represented one level of analysis. Additionally, we look at one exemplary local grassroots organization in each Brandenburg and Ontario. This helps to deepen the analysis and learn more about the organization style and arguments used by the anti-wind movements. Considering two levels of anti-wind protest allows for better grasping the particularity of disputes over wind turbines, which typically start as localized conflicts and are then often up-scaled to the policy-making sub-national level. Interviewees at the state/provincial level included speakers of the anti-wind umbrella organizations, key policy makers, consultants, advocates for renewable energy, municipal organizations and wind energy companies. The local level interviewees encompassed local grassroots anti-wind organizations, regional and township planners, municipal representatives and locally active wind energy companies. Data analysis was informed by the argumentative discourse analysis approach (Hajer 2006) and was practically guided by the coding process of Grounded Theory (Strauss and Corbin 1998).

Structure of the book

After a brief introduction, *Chapter 2* presents the conceptual framework of the study. It first discusses two major explanations for movement outcomes drawing from the literature on social movements and then introduces the concept of discursive energy space. The concept draws on discursive-institutionalist accounts and adds concepts of power. The way in which the discursive energy space provides strength to anti-wind movements is discussed by drawing on classical political science accounts of power.

Chapter 3 familiarizes the reader with renewable energy policy and politics in Canada and Germany. It provides an overview of current developments in climate change and energy policies. The chapter then introduces the case study jurisdictions of Brandenburg and Ontario. Both sub-national jurisdictions are forerunners with regard to wind power deployment in their countries, but have faced fierce resistance to it. They serve as examples for jurisdictions who have taken a lead in wind power policy development, but also as jurisdictions who have responded differently to an emerging anti-wind movement. The chapter introduces the major sub-national policies that governed the jurisdictions' wind power sectors, including the underlying discourses and implementation systems.

Chapters 4 and 5 discuss the rise of the anti-wind movements in Brandenburg and Ontario. The two chapters follow the same structure. This includes a description of how protest manifests itself and what strategies the anti-wind umbrella organizations chose. The section also discusses the way in which the major opposition party perceived and reacted to the anti-wind movement. Furthermore, the response of the Brandenburg and Ontario government to the anti-wind movement is presented. Subsequently, one case study each of a local conflict over wind turbines is discussed. This includes the views, strategies and characteristics of the local conflict.

Chapter 6 first compares the anti-wind movements of Brandenburg and Ontario and the response they received from the government. The chapter then applies the concept of discursive energy space to the empirical findings and shows how the concept is helpful for analyzing the discursive-institutional context when investigating different governmental responses to anti-wind movements. It also spells out which power effects the discursive energy space had in the case studies and how this affected the course of conflict.

Chapter 7 sums up the theoretical argument and discusses the main lessons of the book in light of the recent new climate movement. It draws attention to three important topics that will be decisive for the future development of wind power and other decarbonization projects. First of all, the further development will depend on the way in which these projects are carried out and which participation opportunities for the local population exist. Second, populist forces have evolved who are changing the terms of discussion and thereby seriously challenge the foundation for further decarbonization projects. A third challenge is presented by the incumbent industries who are interested in pursuing their business model. Despite these challenges, the chapter argues, discourses on decarbonization projects will ultimately change considerably as climate change impacts become more apparent in the near future.

References

Agentur für Erneuerbare Energien (2012): Brandenburg – Gesamtsieger Leitstern 2012. Berlin. Available online at www.unendlich-viel-energie.de/fileadmin/content/Panorama/Veranstaltungen/LEITSTERN_2012/Factsheets/AEE_LEITSTERN2012_Brandenburg.pdf, updated on 12/12/2012, checked on 3/12/2013.

Alt, Franz; Claus, Jürgen; Scheer, Hermann (Eds.) (1998): Windiger Protest. Konflikte um das Zukunftspotential der Windkraft. Bochum: Ponte-Press.

Aydemir, Nermin; Vliegenthart, Rens (2017): Public Discourse on Minorities: How Discursive Opportunities Shape Representative Patterns in the Netherlands and the UK. In *Nationalities Papers* 4 (2), pp. 1–15. DOI: 10.1080/00905992.2017.1342077.

Burningham, Kate (2000): Using the Language of NIMBY: A Topic for Research, Not an Activity for Researchers. In *Local Environment* 5 (1), pp. 55–67. DOI: 10.1080/135498300113264.

Cowell, Richard; Bristow, Gill; Munday, Max (2011): Acceptance, Acceptability and Environmental Justice: The Role of Community Benefits in Wind Energy Development. In *Journal of Environmental Planning and Management* 54 (4), pp. 539–557. DOI: 10.1080/09640568.2010.521047.

del Río, Pablo; Linares, Pedro (2014): Back to the Future? Rethinking Auctions for Renewable Electricity Support. In *Renewable and Sustainable Energy Reviews* 35, pp. 42–56. DOI: 10.1016/j.rser.2014.03.039.

Devine-Wright, Patrick (2005): Local Aspects of UK Renewable Energy Development. Exploring Public Beliefs and Policy Implications. In *Local Environment* 10 (1), pp. 57–69. DOI: 10.1080/1354983042000309315.

Devine-Wright, Patrick (2009): Rethinking NIMBYism: The Role of Place Attachment and Place Identity in Explaining Place-Protective Action. In *Journal of Community and Applied Social Psychology* 19 (6), pp. 426–441. DOI: 10.1002/casp.1004.

Dryzek, John S. (1997): The Politics of the Earth. Environmental Discourses. Oxford, New York: Oxford University Press.

Eisinger, Peter K. (1973): The Conditions of Protest Behavior in American Cities. In *The American Political Science Review* 67 (1), pp. 11–28.

Ellis, Geraint; Ferraro, Gianluca (2016): The Social Acceptance of Wind Energy. Where we Stand and the Path Ahead. EUR 28182 EN. European Atomic Energy Community. Available online at http://publications.jrc.ec.europa.eu/repository/bitstream/JRC103743/jrc103743_2016.7095_src_en_social%20acceptance%20of%20wind_am%20-%20gf%20final.pdf, checked on 12/28/2017.

Feindt, Peter H.; Oels, Angela (2005): Does Discourse Matter? Discourse Analysis in Environmental Policy Making. In *Journal of Environmental Policy & Planning* 7 (3), pp. 161–173. DOI: 10.1080/15239080500339638.

Flinders, Matthew; Buller, Jim (2006): Depoliticisation: Principles, Tactics and Tools. In *British Politics* 1 (3), pp. 293–318. DOI: 10.1057/palgrave.bp.4200016.

Flyvbjerg, Bent (2011): Chapter 17. Case Study. In Norman K. Denzin, Yvonna S. Lincoln (Eds.): *The Sage Handbook of Qualitative Research.* 4th ed. Thousand Oaks: Sage, pp. 301–316.

Foucault, Michel (2000): Truth and Power. In Michel Foucault, Paul Rabinow, James D. Faubion (Eds.): *The Essential Works of Michel Foucault, 1954–1984.* New York: New Press.

Gamson, William A. (1975): The Strategy of Social Protest. Homewood Ill.: Dorsey Press (The Dorsey series in sociology).

Goedkoop, Fleur; Devine-Wright, Patrick (2016): Partnership or Placation? The Role of Trust and Justice in the Shared Ownership of Renewable Energy Projects. In *Energy Research & Social Science* 17, pp. 135–146. DOI: 10.1016/j.erss.2016.04.021.

Government of Canada (2016): Pan-Canadian Framework on Clean Growth and Climate Change. Canada's plan to address climate change and grow the economy. Gatineau, Québec: Environment and Climate Change Canada. Available online at www.canada.ca/content/dam/themes/environment/documents/weather1/20170125-en.pdf, checked on 6/19/2017.

Gross, Catherine (2007): Community Perspectives of Wind Energy in Australia: The Application of a Justice and Community Fairness Framework to Increase Social Acceptance. In *Energy Policy* 35 (5), pp. 2727–2736.

Haggett, Claire; Futak-Campbell, Beatrix (2011): Tilting at Windmills? Using Discourse Analysis to Understand the Attitude-Behaviour Gap in Renewable Energy Conflicts. In *Journal of Mechanisms of Economic Regulation*, pp. 207–220.

Hajer, Maarten (1993): Discourse Coalitions and the Institutionalization of Practice: the Case of Acid Rain in Great Britain. In Frank Fischer, John Forester (Eds.): The ArgumentativeTurn in Policy Analysis and Planning. Durham, NC: Duke University Press, pp. 43–76.

Hajer, Maarten (1995): The Politics of Environmental Discourse. Ecological Modernization and the Policy Process. Oxford: Clarendon Press.

Hajer, Maarten (2006): Doing Discourse Analysis. Coalitions, Practices, Meaning. In Brink, Margo van den (Ed.): Words Matter in Policy and Planning. Discourse Theory and Method in the Social Sciences. Utrecht: Koninklijk Nederlands Aardrijkskundig Genootschap (Netherlands geographical studies, 344), pp. 65–76. Available online at www.maartenhajer.nl/images/stories/20080204_MH_wordsmatter_ch4.pdf, checked on 12/1/2014.

Hajer, Maarten; Versteeg, Wytske (2005): A Decade of Discourse Analysis of Environmental Politics: Achievements, Challenges, Perspectives. In *Journal of Environmental Policy & Planning* 7 (3), pp. 175–184. DOI: 10.1080/15239080500339646.

Hantrais, Linda (1995): Comparative Research Methods. In *Social Research Update* (13), pp. 2–11.

IPCC (2014): Climate Change 2014. Mitigation of Climate Change: Contribution of Working Group III to the Fifth Assessment Report of the Intergovernmental Panel on Climate Change. Cambridge: Cambridge University Press.

IRENA (2017): Rethinking Energy 2017: Accelerating the Global Energy Transformation. Edited by International Renewable Energy Agency. Abu Dhabi.

IRENA (2019): Renewable Energy Highlights. Edited by International Renewable Energy Agency. Abu Dhabi.

IRENA; IEA; REN21 (2018): Renewable Energy Policies in a Time of Transition. Available online at www.irena.org/publications/2018/Apr/Renewable-energy-policies-in-a-time-of-transition, checked on 10/11/2019.

Jänicke, Martin; Schreurs, Miranda; Töpfer, Klaus (2015): The Potential of Multi-Level Global Climate Governance. IASS Policy Brief 2/2015. Institute for Advanced Sustainability Studies (IASS). Potsdam.

Jobert, Arthur; Laborgne, Pia; Mimler, Solveig (2007): Local Acceptance of Wind Energy: Factors of Success Identified in French and German Case Studies. In *Energy Policy* 35 (5), pp. 2751–2760.

Kitschelt, Herbert P. (1986): Political Opportunity Structures and Political Protest: Anti-Nuclear Movements in Four Democracies. In *British Journal of Political Science* 16 (1), pp. 57–85.

Koopmans, Ruud; Olzak, Susan (2004): Discursive Opportunities and the Evolution of Right-Wing Violence in Germany. In *American Journal of Sociology* 110 (1), pp. 198–230. DOI: 10.1086/386271.

Koopmans, Ruud; Statham, Paul (1999): Ethnic and Civic Conceptions of Nationhood and the Differential Success of the Extreme Right in Germany and Italy. In Marco Giugni (Ed.): How Social Movements Matter. Minneapolis, MA: University of Minnesota Press (Social movements, protest, and contention, 10), 225–252.

Krogh, Carmen M. E. (2011): Industrial Wind Turbine Development and Loss of Social Justice? In *Bulletin of Science, Technology & Society* 31 (4), pp. 321–333.

Kwon, Tae-hyeong (2015): Rent and Rent-Seeking in Renewable Energy Support Policies. Feed-In Tariff vs. Renewable Portfolio Standard. In *Renewable and Sustainable Energy Reviews* 44, pp. 676–681. DOI: 10.1016/j.rser.2015.01.036.

Lacal Arantegui, Roberto; Jäger-Waldau, Arnulf (2018): Photovoltaics and Wind Status in the European Union after the Paris Agreement. In *Renewable and Sustainable Energy Reviews* 81, pp. 2460–2471. DOI: 10.1016/j.rser.2017.06.052.

Lauber, Volkmar; Schenner, Elisa (2011): The Struggle over Support Schemes for Renewable Electricity in the European Union. A Discursive-Institutionalist Analysis. In *Environmental Politics* 20 (4), pp. 508–527. DOI: 10.1080/09644016.2011.589578.

Leipold, Sina; Feindt, Peter H.; Winkel, Georg; Keller, Reiner (2019): Discourse Analysis of Environmental Policy Revisited: Traditions, Trends, Perspectives. In *Journal of Environmental Policy & Planning* 21 (5), pp. 445–463. DOI: 10.1080/1523908X.2019.1660462.

Leung, Dennis Y.C.; Yang, Yuan (2012): Wind Energy Development and its Environmental Impact: A Review. In *Renewable and Sustainable Energy Reviews* 16 (1), pp. 1031–1039. DOI: 10.1016/j.rser.2011.09.024.

Lewis, Joanna I.; Wiser, Ryan H. (2007): Fostering a Renewable Energy Technology Industry. An International Comparison of Wind Industry Policy Support Mechanisms. In *Energy Policy* 35 (3), pp. 1844–1857. DOI: 10.1016/j.enpol.2006.06.005.

Luderer, Gunnar; Pehl, Michaja; Arvesen, Anders; Gibon, Thomas; Bodirsky, Benjamin L.; Boer, Harmen Sytze de et al. (2019): Environmental Co-benefits and Adverse Side-effects of Alternative Power Sector Decarbonization Strategies. In *Nature Communications* 10 (1), p. 5229. DOI: 10.1038/s41467-019-13067-8.

Lukes, Steven (1974): Power. A Radical View. 1. publ. London: Macmillan (Studies in sociology).

MacArthur, Julie. L. (2016): Challenging Public Engagement. Participation, Deliberation and Power in Renewable Energy Policy. In *Journal of Environmental Studies and Sciences* 6 (3), pp. 631–640. DOI: 10.1007/s13412-015-0328-7.

Meyer, David S.; Minkoff, Debra C. (2004): Conceptualizing Political Opportunity. In *Social Forces* 82 (4), pp. 1457–1492. Available online at http://sf.oxfordjournals.org/content/82/4/1457.full.pdf#page=1&view=FitH, checked on 11/2/2016.

Motta, Renata (2015): Transnational Discursive Opportunities and Social Movement Risk Frames Opposing GMOs. In *Social Movement Studies* 14 (5), pp. 576–595. DOI: 10.1080/14742837.2014.947253.

Ottinger, Gwen; Hargrave, Timothy J.; Hopson, Eric (2014): Procedural justice in wind facility siting: Recommendations for state-led siting processes. In *Energy Policy* 65, pp. 662–669.

Rand, Joseph; Hoen, Ben (2017): Thirty years of North American wind energy acceptance research. What have we learned? In *Energy Research & Social Science* 29, pp. 135–148. DOI: 10.1016/j.erss.2017.05.019.

REN21 (2017): Advancing the Global Renewable Energy Transition: Highlights of the REN21 Renewables 2017 Global Status Report in Perspective. Edited by Renewable Energy Policy Network for the 21st Century. Paris.

Rockström, Johan; Gaffney, Owen; Rogelj, Joeri; Meinshausen, Malte; Nakicenovic, Nebojsa; Schellnhuber, Hans Joachim (2017): A Roadmap for Rapid Decarbonization. In *Science (New York, NY)* 355 (6331), pp. 1269–1271. DOI: 10.1126/science.aah3443.

Saidur, R.; Islam; Rahim, N. A.; Solangi, K. H. (2010): A Review on Global Wind Energy Policy. In *Renewable and Sustainable Energy Reviews* 14 (7), pp. 1744–1762. DOI: 10.1016/j.rser.2010.03.007.

Schmidt, Vivien A. (2008): Discursive Institutionalism: The Explanatory Power of Ideas and Discourse. In *Annual Review of Political Science* 11 (1), pp. 303–326. DOI: 10.1146/annurev.polisci.11.060606.135342.

Schmidt, Vivien A. (2011): Reconciling Ideas and Institutions Through Discursive Institutionalism. In Daniel Béland, Robert Henry Cox (Eds.): Ideas and Politics in Social Science Research. Oxford, New York: Oxford University Press, pp. 47–64.

Schmidt, Vivien A. (2012): Discursive Institutionalism: Scope, Dynamics, and Philosophical Underpinnings. In Frank Fischer, H. Gottweis (Eds.): The Argumentative Turn Revisited: Public Policy as Communicative Practice. Durham, London: Duke University Press, pp. 85–113.

Schmidt, Vivien A. (2015): Discursive Institutionalism: Understanding Policy in Context. In Frank Fischer, Douglas Torgerson, Anna Durnová, Michael Orsini (Eds.): Handbook of Critical Policy Studies. Cheltenham, Northampton, MA: Edward Elgar Publishing (Handbooks of research on public policy), pp. 171–189.

Simcock, Neil (2016): Procedural Justice and the Implementation of Community Wind Energy Projects. A case study from South Yorkshire, UK. In *Land Use Policy* 59, pp. 467–477. DOI: 10.1016/j.landusepol.2016.08.034.

Stokes, Leah C. (2013): The Politics of Renewable Energy Policies: The Case of Feed-In Tariffs in Ontario, Canada. In *Energy Policy* 56, pp. 490–500. DOI: 10.1016/j.enpol.2013.01.009.

Strauss, Anselm L.; Corbin, Juliet M. (1998): Basics of Qualitative research. Techniques and Procedures for Developing Grounded Theory. 2nd ed. Thousand Oaks: Sage Publications.

Szarka, Joseph (2004): Wind Power, Discourse Coalitions and Climate Change: Breaking the Stalemate? In *European Environment* 14 (6), pp. 317–330.

Thygesen, Janne; Agarwal, Abhishek (2014): Key Criteria for Sustainable Wind Energy Planning – Lessons from an Institutional Perspective on the Impact Assessment Literature. In *Renewable and Sustainable Energy Reviews* 39, pp. 1012–1023. DOI: 10.1016/j.rser.2014.07.173.

Valentine, Scott V. (2015): Wind Power Politics and Policy. Oxford: Oxford University Press.

Veers, Paul; Dykes, Katherine; Lantz, Eric; Barth, Stephan; Bottasso, Carlo L.; Carlson, Ola et al. (2019): Grand Challenges in the Science of Wind Energy. In *Science* 366 (6464). DOI: 10.1126/science.aau2027.

Verbruggen, Aviel; Lauber, Volkmar (2012): Assessing the Performance of Renewable Electricity Support Instruments. In *Energy Policy* 45, pp. 635–644. DOI: 10.1016/j. enpol.2012.03.014.

Vrablikova, Katerina (2014): How Context Matters? Mobilization, Political Opportunity Structures, and Nonelectoral Political Participation in Old and New Democracies. In *Comparative Political Studies* 47 (2), pp. 203–229. DOI: 10.1177/0010414013488538.

Wolsink, Maarten (1988): The Social Impact of a Large Wind Turbine. In *Environmental Impact Assessment Review* 8 (4), pp. 323–334. DOI: 10.1016/0195-9255(88)90024-8.

Wolsink, Maarten (2007): Planning of Renewables Schemes: Deliberative and Fair Decision-Making on Landscape Issues Instead of Reproachful Accusations of Non-Cooperation. In *Energy Policy* 35 (5), pp. 2692–2704. DOI: 10.1016/j.enpol.2006.12.002.

Zoellner, Jan; Schweizer-Ries, Petra; Wemheuer, Christin (2008): Public Acceptance of Renewable Energies: Results from Case Studies in Germany. In *Energy Policy* 36 (11), pp. 4136–4141. DOI: 10.1016/j.enpol.2008.06.026.

2 Contentious wind energy and context

Social movements can be characterized by their engagement in conflictual collective action to bring about social change, their dense informal networks and a shared collective identity which goes beyond single events and initiatives (Della Porta and Diani 2006, pp. 20–22). Why are some social movements successful, while others are not? How does the discursive-institutional context against which social movements emerge influence their success? This chapter first conceptualizes government response and looks at how the literature on social movements has analytically approached the importance of context. Then the concept of discursive energy space and its power effects are introduced. The discursive energy space is a framework that identifies important elements to consider when analyzing government response to anti-wind movements. The framework is based on a link between social movement studies, discursive-institutional theories and concepts of power.

Social movements and government response

The government's approach is not necessarily to reject a movement's demands, but may also be to learn from the movement and correspondingly change its policies. The term *government* is used here in the following way. The focus is on the response of the sub-national ruling party or coalition which has the power to change or adapt the contested wind energy policy. Notwithstanding the fact that members of the ruling party or coalition in sub-national governments may adopt different viewpoints regarding the demands of social movements, the focus here is on the statements of the ruling party or coalition of sub-national governments. This includes both responses uttered in written form as well as statements by the energy spokespersons of the ruling parties.

Four different types of responses are distinguished here: integrative, diplomatic, exclusionary and shadow (see Table 2.1). These four types represent a further differentiation of Gamson's classical categorization of movement outcomes as the achievement of *new advantages* and *acceptance* (1975, pp. 28–29). The latter relates to the question of whether a challenging group is accepted "by its antagonists as a valid spokesman for a legitimate set of interests" (Gamson 1975, p. 28), while the former means that the group beneficiaries realize new

Table 2.1 Different types of government response

		Acceptance	
		Recognizing movement as valid and legitimate	Denying legitimacy of movement
New advantages	Adopting demands	**Integrative response**	**Shadow response**
	Declining demands	**Diplomatic response**	**Exclusionary response**

Source: developed from Gamson (1975) and Koopmans and Kriesi (1995).

advantages as a consequence of their actions. New advantages could be the complete or partly adoption of the movement's stated goals. In order to better account for cases in which more differentiated government responses are observable, these two classical types of responses are further distinguished into four types of responses. For this purpose, Koopmans and Kriesi's *integrative* and *exclusive* strategies (1995, pp. 33–34) are added, as well as *shadow* and *diplomatic* response.

An *integrative* government response involves both acceptance and new advantages. If neither acceptance nor new advantages occur, the government response is *exclusionary*. *Exclusionary* strategies include repressive, confrontational, and polarizing responses, while integrative strategies refer to facilitative, cooperative and assimilative responses (Koopmans and Kriesi 1995, pp. 33–34). *Shadow* response is added here to relate to cases when government denies acceptance, but adopts the movement's demands. This may be the case when a government, despite supporting the movement's cause, does not want to be regarded as giving in to the movement or does not want to appear susceptible to blackmail in case the social movement has engaged in violent protest. Finally, *diplomatic* response refers to the situation when a movement is recognized as legitimate and valid, but is not granted new advantages. Governments may adopt this strategy to appear open and responsive to mobilized demands and show a friendly face while not agreeing to the stated demands and thus not adopting the movement's goals. Governments may also agree to the stated demands, but not adopt them because of strategic reasons such as internal party politics. This usually leads to a *diplomatic* response to the movement as well.

A government's response to anti-wind protest is decisive for the fate of a jurisdiction's wind energy policy in the face of protest and is also linked to movement success. Movement success has attracted more attention than government response in the literature on social movements. At first glance, it may seem alluring to simply link movement success to goal achievement as "the unfolding of a process of policy innovation in the political system addressing the protestors' stated needs, if not their actual programs" (Tarrow 1983, p. 5). Due to a number of caveats associated with the concept of movement success, however, the focus here is on government response instead of movement success.

Indeed, "[s]uccess is an elusive idea" (Gamson 1975, p. 28). First, it is not always easy to clearly distinguish between movement "success" and "failure" because different factors such as time or scale have an effect. Governments may for example respond to some demands of the anti-wind movement in an *integrative* way, but *exclusionary* in response to other demands. Demands may also be partly taken over or exhibited in a different way than originally proposed by the movement. Second, anti-wind movements are usually active at two levels. The very local level of individuals protesting against a proposed wind turbine project in their neighborhood may have other demands than their umbrella organization which primarily aims at the political decision-making level to halt the overall wind turbine policy which fosters or supports wind turbine development in one way or the other. It is therefore necessary to differentiate between different levels of the same movement in determining movement success. Governments may respond positively to some demands of the umbrella organization, but not to a local grassroots group demanding the cancellation of a specific project. While a shift in wind energy policies may thus represent a success of the social movement at the upper policy level, it may not be one in the eyes of the local level protest group. Furthermore, outcomes occur along a continuum. Focusing on local-level protest against energy projects, McAdam and Schaffer Boudet differentiate between five local outcomes: "project rejected", "project withdrawn", "approved but withdrawn", "approved but postponed", and "project built" (2012, p. 104). These different categories highlight different possible outcomes of local protest vis-à-vis a planned project.

A third challenge with focusing solely on movement success is the fact that success also refers to those achievements that go beyond directly tangible changes. This makes it difficult to grasp and compare degrees of "success". For example, a movement may not succeed in shaping the public discourse on energy transitions in the short term, but may have an impact over the long run and thereby change policy approaches to renewable energy. Such instances of macrolevel social change are however difficult to trace to specific social movement's actions (McAdam and Schaffer Boudet 2012, p. 102). Macrolevel changes may also be observable only after a given period of time which may make it even more difficult to associate them to a specific social movement. This point relates to the time dimension as the final issue of concern with the concept of movement success. Governments may respond in an *exclusionary* way to anti-wind movements *now*, which could be interpreted as movement failure. However, the protest *now* could turn out to lay the foundation for a more substantial policy response *later*. Due to these difficulties in analyzing movement success, *government response* is the focal point of analysis here rather than *movement success*.

Political and discursive opportunity structure

The analysis of government response necessitates a closer look at the formal institutional system of a polity. As Gillion puts it, the "institutions of government and the individuals who represent it are gatekeepers of social change, and these individuals' willingness to implement reform will be constrained by shifting institutional norms, procedures, and elite alliances" (2013, pp. 35–36). The formal institutional system of a polity does not only relate to the options that social movements have, but also to the options that the government under study has to react to the movement's demands. This institutional system has often been described by the political opportunity structure (POS). The POS was first introduced by Eisinger who defined it as "the degree to which groups are likely to be able to gain access to power and to manipulate the political system" (1973, p. 25). The POS has served as a basic explanatory concept in the study of social movements and has been developed further (e.g. Bloom 2014; Meyer and Minkoff 2004; McAdam 1996; Tarrow 1983; Vrablikova 2014). Kitschelt used POS in his classical comparative work on anti-nuclear power movements in France, Sweden, the United States and West Germany to explain variation in movement strategy and impact on overall energy policy (1986, p. 63). According to Kitschelt, the POS is comprised of the political input and the political output structure, which together fundamentally affect the impact of a movement. The *political input structure* relates to the openness of the political regime to new demands and is made up of the number of political parties, the independence of the legislature in policy development and control, patterns of intermediation between social movement groups and the executive branch and mechanisms for the aggregation of new demands to find their way into policy making processes (Kitschelt 1986, p. 63). The *political output structure*, on the other hand, refers to the capacity of political regimes to implement policies (Kitschelt 1986, p. 64).

However, the POS has been criticized for its narrow focus on the formal institutional structure and for missing elements such as culture, framing, or discourse. The concept of "discursive opportunity structure" (Koopmans and Statham 1999) and more generally, discursive opportunities was therefore introduced to combine the framing perspective with the political opportunity structure. Its main idea is that even though institutional opportunities may be closed, open discursive opportunities may lead to a movement's influence on the public discourse. The role of the media has played a central role for describing discursive opportunities (Aydemir and Vliegenthart 2017; Ferree 2003; Koopmans 2004; Koopmans and Muis 2009; Koopmans and Olzak 2004; Motta 2015). Media representation has therefore been regarded as important for movement outcomes and social movements depend on the media to get their message across (Della Porta and Diani 2006; Gamson 1995; Zald 1996). This understanding suggests that the media has monopoly over and is the gatekeeper of public discourse (Koopmans and Olzak 2004) and that analyzing government response and movement outcomes could be restricted to an analysis of media representation. With the rise of the internet and social media, however, the role of the classic media is changing. Anti-wind movements often have their own websites and social media accounts to spread their message and network with potential supporters. This questions the assumed "opinion monopoly" of the traditional media, despite their important role in framing and shaping public opinion. A further drawback of confining outcome analysis to media analysis is that especially in anti-wind conflicts, it has become a truism that anti-wind protest emerges despite a positive public opinion in favor of renewable energy (Bell et al. 2005; Bell et al. 2013). Media coverage may depict anti-wind protest as positive while opinion polls still show high levels of approval towards renewable energy. This phenomenon suggests that media representation might be overestimated as the exclusive predictor for social movement outcomes and government response.

A discursive approach to anti-wind conflicts

This study takes account of prevailing societal discourses as a central factor, next and in combination with institutional factors, to shape the prospects of social movements to have their issues addressed by the government. Instead of reproducing these overarching discourses by a media content analysis, those discourses will rather be reproduced by a frame analysis (Gamson and Meyer 1996; Johnston 1995, 2002; Noakes and Johnston 2005; Snow 2004) of the anti-wind movements in the case studies and the corresponding reaction by the government. Frames are interpretative schemata and mental constructs, a way of making sense of a complex reality and integrating "facts, values, theories, and interests" (Rein and Schön 1993, p. 145). Frames can be as narrow as applying to one individual only, as broad as reflecting cultural aspects (Johnston and Klandermans 1995; Zald 1996) or can serve as a collective action frame within social movements. A frame shared by a movement inspires and legitimates social movement activities (Snow and Benford 1992). Frame analysis is strongly associated with discourse analysis (Johnston 1995). The way in which policymakers react to anti-wind movements will provide hints on the prevalent discourses that are considered as "legitimate". This is the benefit of a discursive approach to wind energy conflicts, as it stands back from attempts to classify social responses as "valid" or "not valid" but instead,

> it studies how those involved seek to validate their claims and persuade others of their truth, and discredit contradictory claims. This claims making and counter claims making constitutes the debate itself. In this way, the claims are the conflict; there is no other means to access or study it.
>
> (Haggett and Futak-Campbell 2011, p. 18)

A discursive approach is not only helpful with regard to identifying the discursive context. Taking a discursive approach can also help to tackle what Tilly (1999, p. 269) has called the difficulties of identifying social movement outcomes. He posits that no methodological approach will solve the problem of knowing whether an outcome was the result of the movement's actions, the public claims made by movement activists, or by effects outside those events and action. Arguably, many other reasons may be behind governmental decisions to change policies or other institutional arrangements. Those include lobby group activity behind closed doors, internal party politics or personal motivations of decision-makers. In his study on anti-nuclear protest, Kitschelt has also raised the possibility of "autonomous change in the positions, preferences and resources of the nuclear advocates in the four countries during the protest period" (1986, p. 73). Yet, the primary interest here is how the government officially responds to the anti-wind movement and therefore displays the movement in public. The government's framings will be used to decipher whether a change in policy or decision-making system was related to the demands of a social movement or not. If a wind turbine policy is officially changed as a result of "acceptance issues", then it is not necessarily assumed here that this was the only reason to change the policy, but it is acknowledged that the anti-wind movement was at least discursively given enough strength to effect policy change.

Discursive energy space: conceptualizing context

This study argues that the question as to whether a movement's claims enter the policy making arena and possibly lead to *integrative* government responses is a matter of how the existing discursive-institutional system provides discursive strength to the anti-wind movement. The focus is therefore not on either discursive or formal institutional system alone, but on their combined effect. This shall be called the "discursive energy space". Its main elements are the "meaning context" and the "formal institutions". They form the context in which disputes over wind turbines unfold (see Figure 2.1).

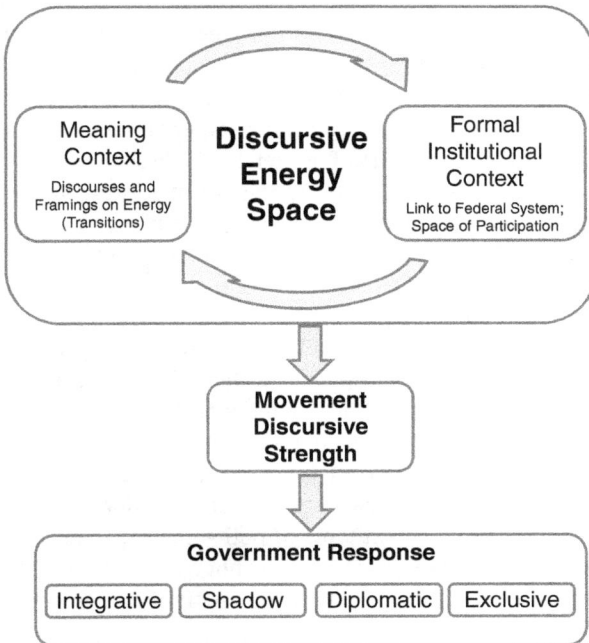

Figure 2.1 The discursive energy space, discursive strength and government response.

The term "discursive energy space" is understood as follows. The term *space* derives from an emphasis on the discursive focus of the study. Environmental policy has been described as "a space where competing narratives of economic and social processes can be revealed" (Ganguly 2013, p. ii). Here, the term *space* is not used to exclusively refer to environmental policy-making, but to describe the sphere of conflict and discursive interaction that takes place within a particular discursive-institutional system of a given jurisdiction. Within this space, the specific focus is on the subject area of *energy*. Energy policy has undergone an enormous degree of change during the last decade, resulting from new discourses informed by rising knowledge on climate change. Framings of the fossil fuel-based energy system have changed, which opened the door to alternative forms of energy sources. Renewable energy has gained importance and many governments around the world follow particular framings of renewable energy to implement corresponding support policies. New discourses on energy have evolved rapidly in the last decade which describes a dynamic *energy space* in which new problem framings are exchanged and debated. In order to delimit this concept of energy space from the study of new physical landscapes of wind energy and in order to set the emphasis on new modes of framings and the evolving disputes over them, *discursive* is added to the term, resulting in *discursive energy space*.

This section provides the theoretical basis for elaborating on the discursive energy space. At the core is a discursive approach which has been used widely in the study of environmental politics (Dryzek 1997, 2013; Feindt and Oels 2005; Hajer 1993; Hajer and Versteeg 2005; Leipold et al. 2019; Szarka 2004). As Kern (2011) notes, these kinds of studies show that new problem framings can lead to political change. This branch of literature on discursive approaches to environmental politics is part of a wider development in policy analysis which focuses on ideas, interests and discourse. Some authors have called it the "argumentative turn" in policy analysis (Fischer 2015; Fischer and Forester 1993; Fischer and Gottweis 2012) and conceive of it as an umbrella term for approaches that stress the importance of ideas, values and discourses in policy making. This branch of literature encompasses various approaches to policy analysis such as framing (Rein and Schön 1993), deliberative approaches (Dryzek and Hendriks 2012) or poststructuralist policy analysis (Howarth and Griggs 2012).

Discourses "enable stories to be told" (Dryzek 2013, p. 17) and their different views on actors and their characteristics, their different assumptions about natural relationships and different perceptions of the world and used rhetorical devices are important elements in discourse analysis (Dryzek 2013, pp. 17–19). Social constructivism (Berger and Luckmann 1966) lies at the heart of this discursive approach, which posits that only discourse confers meaning to a physical or social reality and turns it into a relevant topic of policy interest. This is not to say that anthropogenic climate change for example is not a scientifically observable process, but it only becomes a subject of policy making if it is embedded in a discourse of urgency for action.

As Figure 2.1 shows, the discursive energy space is made up of the meaning context and the formal institutional system. Those will be developed by drawing on three prominent approaches within the discursive school of policy analysis. First, the discursive institutionalism approach by Vivien Schmidt (2008, 2011, 2012, 2015, 2017) will provide the basic understanding on the relationship between discourses and institutions. Second, the interrelated approach of argumentative discourse analysis by Maarten Hajer (1993, 1995, 2006) will add depth to the concepts of discourse, storylines, discourse institutionalization and their link to the framing perspective. Finally, John Dryzek's (2013) account of environmental discourses will be developed to identify four particular energy transition discourses. Despite the fact that all three approaches depart from slightly different starting points, they offer valuable insights into the study of discourses and institutions in environmental policy-making. The three approaches can therefore be regarded as complementary (see also Kern 2011). The next section will discuss them and highlight the way in which they are used here.

Discourses and institutions: setting the scene

The main theoretical approach that this study draws on is discursive institutionalism (DI) as introduced by Schmidt. Discursive institutionalism serves to explain "the dynamics of change (but also continuity) through ideas and discursive interactions" (Schmidt 2011, p. 60). DI is therefore highly relevant for approaching the discursive energy space against which disputes over wind turbines unfold and which leads to different government responses. Studying discursive interaction and discourse development is key to understand a government's decision to respond to an anti-wind movement's demands.

While a range of long-standing traditions in the study of institutions exist in political science (Rhodes 2008), discursive institutionalism (DI) is part of the "new institutionalism" branch of institutional theory. New institutionalism formulates a renewed interest in the study of institutions that aims to be less deterministic, less focused on the pure analysis of formal-legal institutions and more sensitive to the nonpolitical determinants of political behavior than the previous forms of political science institutional analysis (March and Olsen 1984, 2008). Discursive institutionalism is interested in the question of "how are institutions constituted, framed, and transformed through discourse" (Campbell and Pedersen 2001, p. 10). Framing and discourse are thus at the heart of discursive institutionalism. As this study focuses on different framings of wind energy and discourses on energy transition in general, this perspective is particularly relevant.

Discursive institutionalism (DI) is an umbrella term for approaches in political science that focus on institutions while taking "ideas and discourse seriously" (Schmidt 2008, p. 304). Ideas are the "substantive content of discourse" and discourse is regarded as the "interactive process of conveying ideas" (Schmidt 2008, p. 303). Discourse encompasses both the substantive content of ideas but also the communicative process by which they are conveyed. Ideas can be cognitive or normative. Cognitive ideas refer to how policies and programs offer solutions to a problem at hand, while normative ideas instead convey a value-based message to legitimate a particular policy (Schmidt 2008, p. 307). Instead of this notion of ideas, this study refers to "storylines" and "frames" because they better grasp the interpretative and social constructivist dimension of the exchanged arguments. DI defines institutions as both external structures and constructs internal to actors. They are simultaneously conceptualized as "given (as the context within which agents think, speak, and act) and as contingent (as the results of agents' thoughts, words, and actions)" (Schmidt 2008, p. 314).

A major difference to the other new institutionalisms is that DI offers room for conceptualizing institutional change as a direct result of actors' ideas and discourses. Institutional change occurs via "sentient" actors who make use of their "background ideational abilities" and their "foreground discursive abilities" (Schmidt 2008, p. 314). Background ideational abilities relate to an actor's ability to come up with certain ideas and discourses in a certain setting, while foreground discursive abilities refer to an actor's ability to "think and speak outside the institutions in which they continue to act" (Schmidt 2008, p. 315). At the same time, DI also accounts for structural constraints for actors' actions. DI emphasizes both context-dependent aspects *and* purposeful actions as important determinants for institutional change.

This twofold view on institutions as the context within which actors act and as the result of actors' ideas and discourses is of particular relevance for depicting the discursive energy space and accordingly, for explaining government responses to anti-wind opposition. This perspective allows considering the *context* in which disputes over wind turbines occur. For this purpose, Schmidt (2012) has introduced three components of the institutional context: the "meaning context", "formal institutions" and the "forum". This perspective further elucidates the relationship between discourse and institutions. In fact, when a government devises a wind energy policy, it does so by following a particular discourse. The government's ideas on wind energy are conveyed by a discourse which changes the existing energy structure in particular directions, thereby affecting wind energy. Institutional change is therefore observed both at the level of a changing governmental discourse as well as a change in the formal energy policy. Rising protest against wind energy challenges this discourse in an attempt to change it.

In order to elaborate more deeply on this discursive interaction in disputes over wind turbines, we now turn to the argumentative discourse analysis developed by Hajer (1995, 2006). Although Hajer has not explicitly counted himself among the discursive institutionalists, his work can be grasped within the discursive school of policy analysis because of its strong emphasis on how discourse and institutions interrelate. Hajer's work is therefore complementary to Schmidt's work because taking the interrelation between discourse and institutions seriously necessitates an institutional analysis of the "given" institutions along an institutional analysis of the "contingent" idea-type of institutions. This implies a discursive analysis. While the context-part of institutional analysis is thus taken from Schmidt's DI, the discursive analysis part draws on Hajer's work. As a later section will show, these two kinds of perspectives are in fact interrelated.

Hajer defines discourse as "a specific ensemble of ideas, concepts, and categorizations that is produced, reproduced, and transformed in a particular set of practices and through which meaning is given to physical and social realities" (Hajer 1995, p. 264). Hajer argues that political arguments usually draw on several discourses at the same time. Viable policy options for tackling acid rain for example must combine elements of scientific discourse, economical discourse, engineering discourse as well as political considerations (Hajer 1993, p. 46). Illustrating this aspect with the phenomenon of climate change, a policy argument in favor of energy transitions must combine concepts and categorizations from natural science (e.g. using graphs to show rising global temperature), economics (e.g. highlighting costs in the case of non-action) and technology (e.g. showing the workings of a wind turbine). Those sub-set stories of discourses is what Hajer calls "storylines", namely "a generative sort of narrative that allows actors to draw upon various discursive categories to give meaning to specific physical or social phenomena" (Hajer 1995, p. 56). Storylines display different frames which integrate "facts, values, theories and interests" (Rein and Schön 1993, p. 145) and therefore reveal different versions and interpretations of social reality. This is also related to the framing perspective of social movement studies as introduced above. When a group of actors share a set of storylines and therefore "a social construct", they form a discourse coalition (Hajer 1993, p. 45). Discourse coalitions encompasses a set of storylines, the actors that employ them and the practices that relate to these storylines (Hajer 1993, p. 47).

Discourses and institutions interact in a variety of ways. Discourses do not only interpret physical or social realities differently; they also effect institutional change. A discourse can be called dominant if it is translated, or institutionalized, into policies and institutional arrangements (Hajer 1995). When a subnational government is faced with anti-wind opposition and decides to change its wind energy policy or the corresponding decision-making structure, this can be conceptualized as institutional change following discursive interaction.

The meaning context: energy framings and discourses on energy transitions

The first important element of the discursive energy space is the discursive context which shall be grasped here by the first element of Schmidt's "institutional context", i.e. the meaning context. Schmidt's meaning context refers to a setting "in which ideas, arguments and discourses make sense, such that speakers 'get it right' in the ideational rules or rationality of a given setting by addressing their remarks to the 'right' audiences at the 'right' times in the 'right' ways" (Schmidt 2012, p. 105). Schmidt (2012) argues that this bears an important power dimension, as actors may gain power from their ideas, but they may lose power if those do not have sufficient meaning for their audience. For disputes over wind turbines, this implies that the frames used by the anti-wind movement and the government must connect to prevailing discourses in order to be relevant. Those discourses may first of all stem from culture which shapes overall societal discourses linked to the long-term norms and cultural history of a society. Notwithstanding the importance of these macro-discourses, the focus shall here be on the more immediate and volatile discourses and framings that are connected to the issue area of energy and energy transitions. This choice is motivated by the main focus of this study which revolves around the discursive production of a space in which disputes over wind turbines unfold, rather than looking at the broader sociological and cultural aspects of those disputes. The meaning context is therefore conceptualized here as the way in which the energy sector and energy questions are framed in a certain jurisdiction. This may relate to particular topics that the energy sector has long been linked with, but also refers to if and how energy transitions shall take place.

Energy transition discourses are at the heart of conflicts over wind turbines. First of all, they constitute the discursive basis for decisions on wind energy policies. As decision-makers need to legitimize their policy choices, the recurrent reference to particular discourses provides them with a set of arguments and rationales. Furthermore, when anti-wind frames emerge at the local level, those usually also take, directly or indirectly, issue with those energy transition discourses that have informed the contested policy in the first place. As energy transition discourses are the subject of contestation and change, they are more volatile and apt for change than cultural macro-discourses, although they may also be linked to them.

Four energy transition discourses are differentiated here. They draw on John Dryzek's (2013) work on environmental discourses and are further developed for the issue area of energy transitions. Dryzek departs from four basic categories of discourses that have shaped reasoning about the environment not only in the policy-making sphere but in society and industry alike. Those categories include "global limits and their denial", "environmental problem solving", "sustainability" and "green radicalism". While the first category revolves around the acknowledgment that there is the general need to act on environmental problems (or not), the second strand pertains to the way in which those problems shall be tackled. Possible ways to do so are to "leave it to the experts", "to the people" or "to the market" (Dryzek 2013). The discourse category of sustainability includes the discourses on green growth and on ecological modernization, while the discourse category of green radicalism emphasizes the development of green consciousness and consumerism in order to tackle the challenge of environmental protection. Within these four discourse categories, Dryzek identifies eight discourses, out of which four are used here as a starting point for developing energy transition discourses.

Wind energy policies are usually motivated by reference to the discourse on global limits which conveys the finite capacity of planet earth to absorb human waste and pollutants. As harnessing wind resources represents a carbon-neutral source of energy generation, wind turbines are regarded as a relevant technology to mitigate climate change. The following four discourses grasp the major lines of argumentation that can be observed in renewable energy policy making. It is important to note that the discourses presented occur along a scale. Some actors may employ a version of one particular discourse that has similarities with the neighboring discourse.

No need for energy transitions: the fossil discourse

Since industrialization, an economy fueled by fossil resources has been taken for granted. Fossil resources encompass mineral oil, gas and coal, but also uranium is subsumed under this category here. The fossil discourse is based on a belief in abundant resources and unlimited growth. Regarding the long period of an unquestioned use of mineral oil, gas and coal since industrialization, it has only been relatively recently that the fossil discourse has become challenged by a discourse that supports energy transition towards renewable energy. In Germany, the term *"Energiewende"* was first coined by an environmental think tank in the 1980s that called for an energy sector without oil and uranium (Krause et al. 1980). Despite the fact that the fossil discourse is now under pressure by this discursive shift, it still persists as the major challenger to the deployment of renewable energy including wind energy. The fossil discourse does not see any need to change the current system of energy generation. This understanding correlates with Dryzek's "Promethean" discourse. The Promethean position repudiates the need to act on environmental problems and instead believes in human's superiority to nature and denies environmental limits (Dryzek 2013, pp. 52–72). Referring to Greek mythology, Prometheus signifies a fierce confidence that humans and their technologies may overcome any problem.

Adherents to the fossil discourse use various forms of reasoning to dismiss the need for renewable energy. One important sort of reasoning is climate skepticism which comes in at least two disguises (Giddens 2009). The more radical form denies the overall existence of human-induced global warming and believes that observed peaks in temperature are the result of naturally occurring cycles. This understanding is based on a profound denial of scientific findings, nevertheless it is widespread. This position is for example prevalent in the Republican party of the United States. In June 2017, just before President Trump announced the withdrawal from the Paris Agreement, a Republican Congressman said that if there was a real problem resulting from climate change, God would take care of it. This expresses a deep contestation of climate science and also the denial of human capacity to act on it. The other form of climate skepticism acknowledges anthropogenic climate change but believes that its consequences are less relevant than other concerns humanity faces. For adherents of this discourse, poverty or the spread of disease should be paid more attention than combatting climate change by supporting renewable energy. In line with this position, a group within the German Christian-Democratic Party (CDU) called, two days after the US president announced the withdrawal from the Paris Agreement, to direct more attention to climate adaptation rather than mitigation. It was argued that the 2° C target was not achievable and the IPCC's work was dismissed as "moral blackmail". Furthermore, in the September 2017 German federal elections, the *Alternative for Deutschland* was elected to the *Bundestag* as the first political party in Germany that actively puts forward climate skepticism. These examples show that even in a forerunner country such as Germany, the fossil discourse has gained force.

Leaving it to the market: economic rationalism

The economic rationalist discourse concedes the need for solving environmental problems, but argues in favor of market mechanisms to do so (Dryzek 2013, pp. 122–144). Dryzek relates this position to the establishment of property rights, privatization and marketization of environmental goods and related steering mechanisms such as emission trading schemes. This discourse regards humans as economic rational actors and regards market competition as the major mechanism to achieve environmental protection. Economic rationalists would not generally oppose renewable energy – as long as it develops by itself according to market mechanisms. The more conservative proponents of this discourse would reject any governmental subsidy or steering mechanism. This hardline economic rationalist position overlaps with the fossil discourse, as it basically denies the fact that conventional sources of energy generation have once been or are still heavily subsidized, directly or indirectly. The position to reject initial support schemes for renewables is therefore similar to privilege conventional energy sources. There is, however, also a moderate type of the economic rationalist discourse which tends to overlap with the ecological modernization discourse. Adherents of this position are also in favor of market mechanisms and competition, but opt for state-led support schemes to bring about renewable energy capacity. This mainly relates to competitive auction systems in which renewable energy capacity is allotted to the most competitive bidder offering the lowest price. The state plays a steering role in this procurement system, but prices are determined by competition. Auction schemes have been the favorite policy instrument for the introduction of wind energy in North America. As auction schemes require some degree of professionalization in order to be competitive in the bidding procedure, this type of scheme is generally more favorable to larger-scale corporations (Verbruggen and Lauber 2012).

Win/win between economy and environment: ecological modernization

The discourse of ecological modernization emerged in the early 1980s and follows a logic of achieving sustainability by a restructuring of the economic system in a more environmentally friendly direction with a strong role of the state (Dryzek 2013, pp. 165–184). The concept had initially been introduced to reconcile economic with sustainability targets and aimed in its original version at technological renewal beyond end-of-pipe solutions (Jänicke 1984, 2000a). Ecological modernization has later been developed to a more encompassing version including social learning, cultural politics and new institutional arrangements (Hajer 1995; Mol and Spaargaren 2000). Yet, Martin Jänicke, one of the founders of the concept, has pledged to keep the developed version strictly separate from the original concept (Jänicke 2000b). He argues for a strict "technocratic version" of ecological modernization and recognizes that this version will have limitations whenever potentially marketable technical standard solutions are not available (Jänicke 2000a, p. 27). Those relate for example to the unsolved problem of land consumption or biodiversity protection (Jänicke 2000a, p. 27).

Relating the *ecological modernization* discourse to the issue area of energy transitions and renewable energy, two major strands can be distinguished: a technocratic version and a version which is very close to the *democratic pragmatist* discourse described later. The technocratic version of *ecological modernization* has similarities with the *economic rationalist* discourse in that it regards market forces as central to the procurement of renewable energy. Yet, *ecological modernization* emphasizes a strong role of the state. Adherents to the technocratic version of *ecological modernization* would therefore focus on the pure technology procurement part of renewable energy and not on its social dimension. Auction schemes with little focus on local participation would be the favored policy scheme for adherents of technocratic ecological modernization in the realm of renewable energy procurement. On the other end of the discursive range is the version of *ecological modernization* which overlaps with *democratic pragmatism*. Advocates for this version would focus on a renewal of the energy system with a strong public involvement and participation.

From the bottom-up: democratic pragmatism

The "democratic pragmatism" discourse involves the belief that environmental problems should be solved by "leaving it to the people" in the form of a strong role of public participation and deliberation (Dryzek 2013, pp. 99–121). This discourse includes concepts of public consultation, policy dialogue, lay citizen deliberation or public inquiries. Interactive problem solving (Dryzek 2013, p. 99) and the recognition of citizens as valuable and important actors are at the core of this discourse. In the renewable energy sector, a discourse on democratic pragmatism can take several forms. One version finds itself at the intersection with the discourse on ecological modernization and revolves around the steering role of the state to provide platforms for public consultations on wind turbine planning or facilitate local ownership of wind turbines. If this type of discourse is dominant in the policy-making sphere, a possible institutional structure for wind turbine support is a feed-in tariff (FIT) in combination with a meaningful decision-making process for wind turbine siting. A FIT scheme is usually associated with the inclusion of smaller energy producers, such as single households with rooftop photovoltaic or farmers' cooperatives. The description of Germany's success in the deployment of renewables in terms of "energy democracy" (Morris and Jungjohann 2016) is largely attributable to its implementation of a FIT that allowed a large variety of actors to enter the energy sector. This includes community energy which would be the favored organizational form of wind turbine development for democratic pragmatists. This includes co-operative and community-owned types of energy generation (see Walker 2011; Walker and Devine-Wright 2008). There is, however, also a type of democratic pragmatist discourse which sees no role for the state. The hardline democratic pragmatist position would favor wind turbine development in the exclusive hands of the individual – without any state interference. This radical type of democratic pragmatism would most certainly result in less procured capacity than with a supportive role of the state.

Summary and discussion of the four discourses

It is important to note that none of these policy discourses exist in a pure form and are depicted here in their ideal character in order to illustrate their differences in terms of the way they endorse energy transitions. While the *fossil discourse* denies the need to engage in endeavors for energy transitions, the discourses on *market rationalism, ecological modernization* and *democratic pragmatism* differ in the role they attribute to the market, state and the people to procure renewable energy. Policy decisions and disputes over them may follow one of these discourses as their main rationale, but may also borrow aspects from other discourses. As has been shown, discourses may also overlap. Yet, the distinction between discourses is useful because they are also predictive of the very characteristics of and public reaction to energy transitions. For example, the literature on wind energy conflicts suggests that perceived injustice and a lack of community-based initiatives are a key factor in low rates of acceptance. It can therefore be assumed that the *democratic pragmatist discourse*, once institutionalized into specific policies, would meet with less local resistance than the *economic rationalist discourse* which favors large-scale wind farms without a strong role for participation. The *ecological modernization discourse* lies in between, as it attributes a strong role to the state while endorsing energy transitions.

In conclusion, the four forms of energy transition discourses introduced here would lead to different degrees of renewable energy procurement (see Figure 2.2). A dominant *fossil discourse* would most certainly completely prevent the development of wind energy. A dominant *economic rationalist* discourse would do better in this regard, but as its emphasis is on large-scale developments without considerable public input, it would probably be less efficient in procuring wind energy capacity than a dominant *ecological modernization* or *democratic pragmatism* discourse. The *democratic pragmatism* discourse embraces local ownership of wind turbines which are embedded in grassroots support for wind energy. As has been shown for Denmark, this type of wind energy has resulted in a larger number of turbines as compared to the UK where large companies and the financial sector – typical beneficiaries of the economic rationalist discourse – dominate the sector (Toke 2002). As the democratic pragmatist discourse involves a high level of decentralization and public involvement, it can therefore be assumed that this kind of decision-making and benefit sharing by the public leads to high realization rates of renewable energy projects.

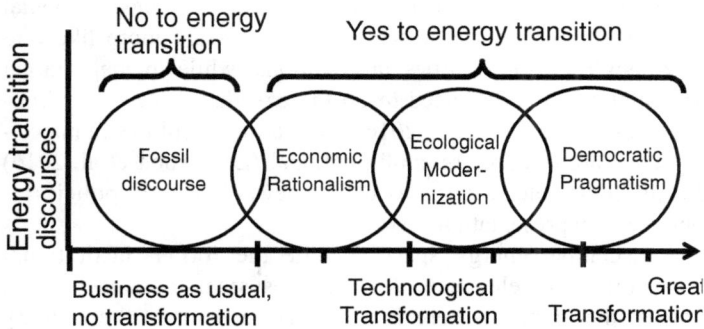

Figure 2.2 Discourses on energy transitions.
Source: developed from Dryzek 2013.

Summing up, Figure 2.2 illustrates how the fossil discourse endorses a "business as usual" scenario that does not lead to any energy transition. Economic rationalism and ecological modernization advocate energy transition, which can be understood as "technological transformation", while the democratic pragmatist discourse stands for a "great transformation". This type of energy transition does not only imply a technological shift in energy systems, but a change of society as a whole. It includes a re-structuring of the economy and living conditions under maximum citizen participation and ownership (WBGU 2011).

The formal institutional context

The second element of the discursive energy space is the formal institutional context. This corresponds, but goes beyond, Schmidt's conception of *formal institutions* which "affect *where* discourse matters by establishing who talks to whom about what, where, and when" (Schmidt 2012, p. 106, italics in original). Schmidt mainly relates this to the observation that countries with a proportional representation system such as Germany tend to have more elaborate *coordinative* discourse about policy construction among policy actors (Schmidt 2008). In majoritarian polities such as the United States, a *communicative* discourse with the public about the necessity and appropriateness of policies is more pronounced. The make-up of the formal political system can therefore have an important impact on how policy ideas are negotiated and contested at the policy-making level and how discourses on energy transitions are put forward. Kern (2011) for example shows in a comparative study that the consensus-oriented political system in the Netherlands and the majoritarian governance system of the UK account for differences in the way in which policy initiatives for sustainability transitions are undertaken. For the case of the Netherlands, Kern observes that any new policy approach has to be linked to dominant discourses of the time (the meaning context) in order to be successful, while in the UK, the government can create "new policies without too many checks and balances" (Kern 2011, p. 1126). The electoral system can also have an effect on how environmental topics come up at all. Proportional representation systems are more likely to include smaller pro-environmental parties in coalitions, while in majoritarian systems, environmental NGOs often need to do the agenda-setting of environmental policy measures (Lockwood et al. 2016). Furthermore, politics in majoritarian electoral systems are often more confrontational (Lockwood et al. 2016) which may add a different note to disputes over renewable energy policies as compared to proportional representation systems.

How does the discursive energy space include the formal institutional setting? Notwithstanding the relevance of electoral system to impact social movement outcomes, the emphasis here is a discursive one. As the primary interest is on the government response to anti-wind protest and the discursive interaction within this dispute, the focus is here on how the formal institutional system contributes to movement discursive strength to successfully scale-up the movement's concerns to the policy-making level. This is connected to the question how the contested sub-national wind energy policy is embedded in the federal system and how this impacts movement strength. If a sub-national wind energy policy is firmly embedded in federal energy policy, it can add discursive strength to the sub-national government to refer protest to the federal level instead of dealing with it at the sub-national level. One part of the formal institutional context of the discursive energy space shall therefore be the degree to which the contested sub-national wind energy policy is part of federal energy policy.

The second part of the formal institutional context of the discursive energy space is the "space of participation". This draws on Schmidt's third element of the institutional context, i.e. the *forum*. The forum depicts an arena of argumentation defined by certain procedural rules or a "particular logic of communication" (Schmidt 2012, p. 105). Schmidt gives the example of a courtroom or international negotiations that may take place in ad-hoc negotiations because pre-established rules do not exist. In contrast to Schmidt, the *local* dimension of the formal institutional system for wind energy disputes is addressed here. Disputes over wind energy usually emanate at the very local level when residents oppose a proposed or actual development of wind turbines in their area. If the concerned residents could prevent the project from being built already at the local level, the conflict would probably not reach the upper policy level. As the literature on anti-wind disputes suggests, the local decision-making for wind turbine development is key in wind turbine conflicts. Wolsink for instance argues that "the institutional setting of the 'decision-making level' is highly significant for the aggregate outcome of wind power implementation at the national level" (2007, p. 2694) because institutional decision-making systems that work in a top-down manner and do not incorporate local values lead to opposition and may turn qualified supporters into opponents. In this sense, opposition to wind turbines evolving from the decision-making system may feed back into the policy-making level.

This local level of the formal institutional context as included here refers to the decision-making or planning system for wind turbines as "spaces of participation" (Gaventa 2006). They are endorsed, changed or newly created by a prevailing discourse on energy transitions. Those spaces can be regarded as "opportunities, moments and channels where citizens can act to potentially affect policies, discourses, decisions and relationships that affect their lives and interests" (Gaventa 2006, p. 26). They represent the place in which local storylines for and against wind turbines unfold and where local opposition becomes apparent. As the presentation of the meaning context has shown, different discourses on energy transitions attribute importance to the space of participation. While the *economic rationalism* discourse focusing on market solutions for renewable energy procurement would only endorse a very restricted space of participation if any at all, the *democratic pragmatist* discourse would embrace an extensive space of participation.

The power of anti-wind movements

This section takes a closer look at how the discursive energy space precisely impacts government response to anti-wind protest. The main argument is, as spelled out before, that the characteristics of the discursive energy space may increase or weaken the movement's discursive strength and therefore prompt the government to respond in one way or the other. Anti-wind storylines can ultimately challenge overarching policy discourse in various directions, i.e. from *economic rationalist* to *democratic pragmatist* or to the *fossil discourse*. Disputes over wind turbines then become a "struggle for discursive hegemony" (Hajer 1995, p. 59). This conceptualization of the relationship between discursive strength and government response is based on an understanding of government decision-making as a negotiation process. A government faced with anti-wind opposition will carefully consider the arguments put forward and decide about its response. These negotiation processes, however, unfold against the background of powerful discursive and institutional structures which are described by the discursive energy space. The concept is therefore closely connected to the political science concept of power.

Classical approaches to power

Power is one of the most important concepts in political science. Approaches, definitions and classifications abound.[1] As power is an "essentially contested concept" (Lukes 1974), it has been argued that the use of such a loaded term may be normative, evaluative or analytical (Haugaard and Ryan 2012a). Clearly, analyzing how a discursive energy space provides "power" to the anti-wind movement runs the risk of following an agenda that either dismisses the legitimacy of the anti-wind movement or the government's approach to dealing with it. The attempt here is, however, to use the concept of power as an analytical tool to understand how the anti-wind movement gains or loses strength to have an effect on the government. For this purpose, four classical conceptions of power are presented here. They include the three dimensions of power proposed by Steven Lukes (1974) and a fourth dimension which draws on the work of Michel Foucault (2000). Acknowledging that each of these power conceptions has its own historical background and represents an extensive theoretical strand of debate, they are presented and used here in a strongly simplified manner in order to highlight their possible contribution to the purpose of this study. Those four faces of power will now be briefly introduced and discussed with regard to their relevance for the effects of the discursive energy space to provide discursive strength to the anti-wind movement.

Lukes (1974) argues that power should be studied by looking at its three faces. His argument is embedded in a larger debate over the notion of classical pluralism that everyone has an equal amount of power no matter their social status. The first face relates to "decision-making power". It grasps power as a behavioral attribute and draws on Dahl's pluralist formula of power. In a conflict of interest, A has power over B to the extent that A can "get B to do something that B would not do otherwise" (1957, p. 201). This account of power can also be applied to situations of brute force or coercion and compulsion, but the account of decision-making is more relevant here. Most importantly, this notion of power will be used here for describing "the power of the anti-wind movement". If a social movement successfully pressures the government to act on their demands and thus prompts an *integrative* government response, the movement "has power" in the sense of this first face. This is because the movement may prompt the government to "do something they would not do otherwise", even if the means may include persuasion or other soft forms of pressure instead of coercion or brute force in the classical sense of the first face of power. This first form of power thus plays a role at the impact level of social movements.

The second face of power is "non-decision-making power" and goes back to Bachrach and Baratz (1962). Responding to Dahl, Bachrach and Baratz proclaimed a two-dimensional concept of power which posits that power relations are also relevant with regard to "non-decisions". Put differently, A may not only play out power by pushing B to a certain decision, but also by devoting their "energies to creating or reinforcing social and political values and institutional practices that limit the scope of the political process to public consideration of only those issues which are comparatively innocuous to A" (Bachrach and Baratz 1962, p. 948). Bachrach and Baratz draw here on Schattschneider's succinct remark:

> All forms of political organization have a bias in favor of the exploitation of some kinds of conflict and the suppression of others because *organization is the mobilization of bias*. Some issues are organized into politics while others are organized out.
>
> (Schattschneider 1960, p. 71, italics in original)

The second face of power thus describes how the context of decision-making is shaped in a way as to exclude some topics from arising in a debate. This may be done consciously or unconsciously and can take discursive or institutional forms. The institutional form relates to the power effects of formal institutions which "shape the context" (Hay 1997) of decision-making. Some institutional systems may for example encourage the discussion of some topics and suppress others. The space of participation in wind turbine siting may for example allow addressing biodiversity-related concerns, but exclude others. If the institutional arrangement is strategically crafted by decision-makers to confine the scope of what can be discussed, they wield this second face of power. This institutionalized form of power has also been called "strategic institutional design" (Landwehr and Böhm 2014) or "institutional depoliticization" (Flinders and Buller 2006). The second face of power is also often used in a discursive way which is closely connected to the institutions in which actors are embedded. Discourse analysis offers the opportunity to expose those power relations in institutional arrangements, unveiling the "creation of structures and fields of action by means of storylines, positioning, and the selective employment of comprehensive discursive systems" (Hajer 1995, p. 275). This discursive use of the second face of power is linked to the fourth face of power and will be further discussed later.

Bachrach and Baratz' view on power was extended by Lukes' "radical view" (1974) which introduces a third face to the notion of power. In short, Lukes refers to the possibility that A exercises power over B by shaping B's preferences which B is not aware of. This form of power can be regarded as covert manipulation. Analyzing this type of power necessitates knowledge over B's "real" preferences in order to delimit them from those imposed by A. This poses a fundamental methodological challenge. According to Hay (1997), it is highly problematic to claim that the researcher investigating this third face of power empirically would exactly know, from an ivory tower perspective, B's "real" preferences. Due to the described methodological challenges, this type of power will not be further pursued in this study.

The three faces of power so far presented assume that power is something that is exerted by actors. Another approach does not locate power within individuals but regards it as ubiquitous. This poststructuralist notion of power can be called a fourth face of power (e.g. Partzsch 2015) and draws on the work of Michel Foucault. Emphasizing the overall prevalence of power, Foucault stresses the societal and universal power aspect of discourse:

> Each society has its régime of truth, its 'general polities' of truth: that is, the types of discourse which it accepts and makes function as true; the mechanisms and instances which enable one to distinguish true and false statements.
>
> (2000, p. 131)

According to this school of thought, prevailing discourses in a society determine what is deemed as true and what is not. They contribute to the "prevailing definition of reality" and can be called "powerful" in terms of shaping the context of societal interaction. Poststructuralist policy analysis therefore stresses "the practices of policy making in relation to wider social and political contexts" (Howarth and Griggs 2012, p. 310). While Foucault developed this concept of power rather in relation to longer-term historic periods in societies, his conceptions can also be applied to characterize the workings of power in medium-term energy transitions (Bues and Gailing 2016). The concept has also informed social movement studies (Baumgarten and Ullrich 2016). The German discourse on the importance of phasing out nuclear energy for example (the discourse on *Energiewende*) can be counted as such a powerful discourse of this fourth face of power. As discourses form an important part of the meaning context, the fourth face of power is particularly relevant.

The fourth face of power is also relevant in argumentative discourse analysis, which aims at unveiling how "power is structured in institutional arrangements" (Hajer 1995, p. 264). In Hajer's understanding, the "politics of discourse is (. . .) about the actual creation of structures and fields of action by means of storylines, positioning, and the selective employment of comprehensive discursive systems" (Hajer 1995, p. 275). He thus stresses the power dimension of discourses that are intentionally put forward by actors to create institutions, but he also acknowledges that actors are constrained by context such as "conventional understandings and agreed-upon rules of the game" (Hajer 1995, p. 275). Hajer's understanding of power therefore mainly relates to the creation of institutions by discourse, the power effects of those newly created institutions, but also to the power effects of existing context.

In light of these considerations, a framework is needed for discursive-institutional analysis that combines the power effects of discursive and institutional elements on an actor's power to influence or persuade others to adopt their ideas or positions. The next section therefore introduces the power concept of the discursive energy space, referring to the classical political science concepts of power, i.e. the first, second, and fourth faces of power.

Discursive strength and movement power

The peculiarity of disputes over wind turbines is that they occur at different levels. Usually, anti-wind conflicts first become visible as a localized dispute over the development of wind turbines in a specific area. Only if the discursive framings and concerns reach the government level, do they become policy-relevant in that they may change existing wind power policies. This is the case if a social movement is accepted "by its antagonists as a valid spokesman for a legitimate set of interests" (Gamson 1975, p. 28). A movement's discursive strength can thus be defined as its ability to scale-up its concerns to the sub-national government. This is achieved when the movement is considered as a legitimate spokesperson for a legitimate set of interests.

Various power effects of the discursive energy space can be identified that influence a movement's discursive strength (see Table 2.2). The meaning context impacts a movement's ability to scale-up its concerns to the policy-making level in two major ways. First, the existence of powerful discourses may make a

Table 2.2 The power dimension of the discursive energy space

Discursive energy space	Relevant face of power	Effect on the ability of anti-wind movements to scale-up their concerns ("discursive strength")
Meaning context	4th face of power	Existence of powerful discourses make a particular discourse on wind turbines more valid than others → weakening or reinforcing movement discursive strength.
	2nd face of power	Those confronted with anti-wind opposition may refer to particular discourse to reject demands → weakening movement discursive strength.
Institutional context: link to federal energy policy	2nd face of power	Those confronted with anti-wind opposition may refer to the federal level to reject addressing the movement's demands → weakening movement discursive strength.
Institutional context: space of participation	1st face of power	The space of participation may restrict participation directly (1st face) or by reducing the concerns that are deemed valid (2nd face) → weakening movement discursive strength.
	2nd face of power	Those confronted with anti-wind opposition may refer to space of participation to reject demands → weakening movement discursive strength.

particular framing used by the anti-wind movement more valid than others. If the concerns of an anti-wind movement for example correspond to an energy transition discourse which is strongly represented in society in general, the movement gains discursive strength. This is related to the fourth face of power which entails that there are some types of discourses in a society "which it accepts and makes function as true" (Foucault 2000, p. 131), while other discourses in other societies may be regarded as less valid. If the particular wind energy policy which an anti-wind movement targets is generally in line with such an accepted and thus powerful discourse in a society, it may weaken a movement's discursive strength.

The second and related way in which the meaning context impacts a movement's discursive strength is via the second face of power. Prevailing discourses may not only indirectly make a movement's claims "more" or "less" valid, but the government which is confronted with the movement may also directly refer to discourses prevailing in the meaning context to reject the movement's demands. If for instance the *ecological modernization* discourse is very strong in a given meaning context, the government may reject the movement's demands to stop wind turbine development by referring to the need to modernize the economy. This kind of non-decision power may be deliberately used to "limit the scope of the political process to public consideration of only those issues which are comparatively innocuous to A" (Bachrach and Baratz 1962, p. 948). An anti-wind movement can therefore gain discursive strength if their ideas, framings and stories on social reality resonate with these prevalent discourses. This is exemplified by Vivien Schmidt's statement that "actors can gain power from their ideas even where they may lack the power of position" (Schmidt 2012, p. 106). This understanding of discursive strength depicted here is hence a development and specification of Schmidt's earlier work.

In a similar vein, the formal institutional context made up of the degree to which the contested energy policy is embedded in federal energy policy and the space of participation also has power effects and impacts discursive strength. If a contested sub-national wind energy policy is part of a federal energy policy, sub-national decision-makers may make use of the second face of power and thereby lower a movement's possibility to scale-up their concerns. This is because sub-national members of government may refer to the federal level of politics to deal with the concerns raised by the anti-wind movement. They thereby "depoliti-cize" (Flinders and Buller 2006) the concerns and therefore make it more diffi-cult for the anti-wind movement to be positively responded to at the sub-national level.[2] This power effect of the institutional context refers to the second face of power as it limits the "scope of the political process" (Bachrach and Baratz 1962, p. 948) at the sub-national level.

Finally, the space of participation can directly restrict the participation of oppon-ents (the first face of power) as well as the arguments that are regarded as legitimate to be put forward by those contesting wind turbines (the second face of power). This is for example the case when formal participation processes of wind turbine planning only allow arguments relating to biodiversity, but not concerns for land-scape aesthetics. This restriction of accepted arguments impacts the ability of the anti-wind movement to have their specific concerns addressed and up-scaled to the policy-making level via the space of participation. The more direct and obvious this first face of power is played out via the space of participation, the heavier a reaction from anti-wind movements may be. This may also add to a movement's discursive strength. Moreover, the space of participation may also have power effects in the sense of the second face of power. When a government is confronted with particular demands of the anti-wind movement, it may hint at the responsibility of the space of participation to deal with this kind of demand. This reference to another political level to address the movement's concerns limits the "scope of the political process" (Bachrach and Baratz 1962, p. 948) at the sub-national government level and there-fore weakens a movement's discursive strength.

To sum up, the discursive energy space introduced in this section matters with regard to the ability of the anti-wind movement to scale-up its concerns. While certainly all forms of power can be identified as relevant for the discursive energy space, the emphasis was here on the first, second and fourth face of power. They are particularly identified here as impacting a movement's discur-sive strength. Applying the framework of discursive energy space to the study of different government responses to anti-wind protests in the sub-national case study jurisdictions of Brandenburg and Ontario has several implications. First, it

Table 2.3 Operationalization of discursive energy space for empirical analysis

Theoretical components of the discursive energy space	Guiding questions for empirical analysis
Meaning context	Which discourses are behind the government's wind energy policy? Which discourses does the anti-wind movement use to oppose wind energy development?
	How does the government explain its reaction to the anti-wind movement? Which concerns does the government accept and which ones does it not accept?
Formal institutional context: link to federal energy policy	How is the sub-national government's energy policy embedded in the federal energy policy?
	Does the government refer to the federal level when responding to the anti-wind movement?
Formal institutional context: space of participation	How is the government's energy policy implemented in the space of participation? How does the government frame the space of participation? How does the anti-wind movement frame the space of participation? Does the government change the space of participation in reaction to the anti-wind movement?

is necessary to identify the different categories of the framework separately. Second, the interrelation between the elements of the framework need to be assessed. Table 2.3 presents the guiding questions for the empirical analysis of the discursive energy space.

Notes

1 For instance: For the issue area of sustainability research, Partzsch (2015) highlights the common distinctions of "power with", "power to" and "power over". Hearn (2012) distinguishes between the five classifications of "physical versus social power", "power 'to' versus power 'over'", "asymmetrical versus balanced power", "power as structures versus agents", and "actual versus potential power". Haugaard and Ryan (2012b) highlight the "conflictual", the "consensual" and the "constitutive streams" of social and political power.
2 The concepts of power used here are used for analytical and not for normative purposes. The term "depoliticization" may sound judgmental, but is mentioned here as a descriptive concept.

References

Aydemir, Nermin; Vliegenthart, Rens (2017): Public Discourse on Minorities. How Discursive Opportunities Shape Representative Patterns in the Netherlands and the UK. In *Nationalities Papers* 4 (2), pp. 1–15. DOI: 10.1080/00905992.2017.1342077.

Bachrach, Peter; Baratz, Morton S. (1962): Two Faces of Power. In *American Political Science Review* 56 (04), pp. 947–952. DOI: 10.2307/1952796.

Baumgarten, Britta; Ullrich, Peter (2016): Discourse, Power, and Governmentality. Social Movement Research with and beyond Foucault. In Jochen Roose, Hella Dietz (Eds.): Social Theory and Social Movements. Mutual inspirations, pp. 13–38. Wiesbaden: Springer Fachmedien.

Bell, Derek; Gray, Tim; Haggett, Claire (2005): The 'Social Gap' in Wind Farm Siting Decisions: Explanations and Policy Responses. In *Environmental Politics* 14 (4), pp. 460–477.

Bell, Derek; Gray, Tim; Haggett, Claire; Swaffield, Joanne (2013): Re-Visiting the 'Social Gap': Public Opinion and Relations of Power in the Local Politics of Wind Energy. In *Environmental Politics* 22 (1), pp. 115–135. DOI: 10.1080/09644016.2013.755793.

Berger, Peter L.; Luckmann, Thomas (1966): The Social Construction of Reality. A Treatise in the Sociology of Knowledge. Garden City, NY: Doubleday (Anchor books, A589).

Bloom, Jack M. (2014): Political Opportunity Structure, Contentious Social Movements, and State-Based Organizations. The Fight against Solidarity inside the Polish United Workers Party. In *Social Science History* 38 (3–4), pp. 359–388. DOI: 10.1017/ssh.2015.29.

Bues, Andrea; Gailing, Ludger (2016): Energy Transitions and Power: Between Governmentality and Depoliticization. In Ludger Gailing, Timothy Moss (Eds.): *Conceptualizing Germany's Energy Transition*. London: Palgrave Macmillan, pp. 69–91.

Campbell, John L.; Pedersen, Ove K. (2001): The rise of Neoliberalism and Institutional Analysis. Introduction. In John L. Campbell, Ove K. Pedersen (Eds.): *The Rise of Neoliberalism and Institutional Analysis*. Princeton, NJ, Chichester: Princeton University Press.

Dahl, Robert (1957): The Concept of Power. In *Behavioral Science* 2 (3), pp. 201–215.

Della Porta, Donatella; Diani, Mario (2006): Social Movements. An Introduction. 2. ed. Malden Mass: Blackwell.

Dryzek, John S. (1997): The Politics of the Earth. Environmental Discourses. Oxford, New York: Oxford University Press.

Dryzek, John S. (2013): The Politics of the Earth. Environmental Discourses. 3. ed. Oxford: Oxford University Press.

Dryzek, John S.; Hendriks, Carolyn (2012): Fostering Deliberation in the Forum and Beyond. In F. Fischer, H. Gottweis (Eds.): The Argumentative Turn Revisited: Public Policy as Communicative Practice. Durham, London: Duke University Press, pp. 31–57.

Eisinger, Peter K. (1973): The Conditions of Protest Behavior in American Cities. In *The American Political Science Review* 67 (1), pp. 11–28.

Feindt, Peter H.; Oels, Angela (2005): Does Discourse Matter? Discourse Analysis in Environmental Policy Making. In *Journal of Environmental Policy & Planning* 7 (3), pp. 161–173. DOI: 10.1080/15239080500339638.

Ferree, Myra Marx (2003): Resonance and Radicalism: Feminist Framing in the Abortion Debates of the United States and Germany. In *AJS* 109 (2), pp. 304–344.

Fischer, F.; Gottweis, H. (Eds.) (2012): The Argumentative Turn Revisited: Public Policy as Communicative Practice. Durham, London: Duke University Press.

Fischer, Frank (2015): In Pursuit of Usable Knowledge: Critical Policy Analysis and the Argumentative Turn. In Frank Fischer, Douglas Torgerson, Anna Durnová, Michael Orsini (Eds.): Handbook of Critical Policy Studies. Cheltenham, UK, Northampton,

MA, USA: Edward Elgar Publishing (Handbooks of research on public policy), pp. 47–66.

Fischer, Frank; Forester, John (Eds.) (1993): The Argumentative Turn in Policy Analysis and Planning. Durham, NC: Duke University Press.

Flinders, Matthew; Buller, Jim (2006): Depoliticisation: Principles, Tactics and Tools. In *British Politics* 1 (3), pp. 293–318. DOI: 10.1057/palgrave.bp.4200016.

Foucault, Michel (2000): Truth and Power. In Michel Foucault, Paul Rabinow, James D. Faubion (Eds.): The Essential Works of Michel Foucault, 1954–1984. New York: New Press (The essential works of Michel Foucault, 1954–1984, v. 1).

Gamson, William A. (1975): The Strategy of Social Protest. Homewood Ill.: Dorsey Pr (The Dorsey series in sociology).

Gamson, William A. (1995): Constructing Social Protest. In Hank Johnston, Bert Klandermans (Eds.): Social Movements and Culture. Minneapolis, MN: University of Minnesota Press (Social movements, protest, and contention), pp. 85–106.

Gamson, William A.; Meyer, David S. (1996): Framing Political Opportunity. In Doug McAdam, John D. McCarthy, Mayer N. Zald (Eds.): Comparative Perspectives on Social Movements: Political Opportunities, Mobilizing Structures, and Cultural Framings. Cambridge: Cambridge University Press, pp. 275–290.

Ganguly, Sunayana (2013): Deliberating on Envionment Policy in India: Participation and the Role of Advocacy. Dissertation. Freie Universität Berlin, Berlin.

Gaventa, John (2006): Finding the Spaces for Change: A Power Analysis. In *IDS Bulletin* 37 (6), pp. 23–33.

Giddens, Anthony (2009): The Politics of Climate Change. National Responses to the Challenges of Global Warming. Policy Network Paper. London.

Gillion, Daniel Q. (2013): The Political Power of Protest. Minority Activism and Shifts in Public Policy. Cambridge: Cambridge University Press (Cambridge studies in contentious politics).

Haggett, Claire; Futak-Campbell, Beatrix (2011): Tilting at Windmills? Using Discourse Analysis to Understand the Attitude-Behaviour Gap in Renewable Energy Conflicts. In *Journal of Mechanisms of Economic Regulation*, pp. 207–220.

Hajer, Maarten (1993): Discourse Coalitions and the Institutionalization of Practice: The Case of Acid Rain in Great Britain. In Frank Fischer, John Forester (Eds.): The Argumentative Turn in Policy Analysis and Planning. Durham, NC: Duke University Press, pp. 43–76.

Hajer, Maarten (1995): The Politics of Environmental Discourse. Ecological Modernization and the Policy Process. Oxford: Clarendon Press.

Hajer, Maarten (2006): Doing Discourse Analysis. Coalitions, Practices, Meaning. In Brink, Margo van den (Ed.): Words Matter in Policy and Planning. Discourse Theory and Method in the Social Sciences. Utrecht: Koninklijk Nederlands Aardrijkskundig Genootschap (Netherlands geographical studies, 344), pp. 65–76. Available online at www.maartenhajer. nl/images/stories/20080204_MH_wordsmatter_ch4.pdf, checked on 12/1/2014.

Hajer, Maarten; Versteeg, Wytske (2005): A Decade of Discourse Analysis of Environmental Politics: Achievements, Challenges, Perspectives. In *Journal of Environmental Policy & Planning* 7 (3), pp. 175–184. DOI: 10.1080/15239080500339646.

Haugaard, Mark; Ryan, Kevin (2012a): Introduction. In Mark Haugaard, Kevin Ryan (Eds.): Political Power. The Development of the Field. Opladen: Barbara Budrich (World of political science: The development of the discipline book series), pp. 9–20.

Haugaard, Mark; Ryan, Kevin (2012b): Social and Political Power. In Mark Haugaard, Kevin Ryan (Eds.): Political Power. The Development of the Field. Opladen: Barbara

Budrich (World of political science: The development of the discipline book series), pp. 21–54.

Hay, C. (1997): Divided by a Common Language. Political Theory and the Concept of Power. In *Politics* 17 (1), pp. 45–52. DOI: 10.1111/1467-9256.00033.

Hearn, Jonathan (2012): Theorizing Power. Basingstoke: Palgrave Macmillan.

Howarth, David; Griggs, Steven (2012): Poststructuralist Policy Analysis: Discourse, Hegemony, and Critical Explanation. In F. Fischer, H. Gottweis (Eds.): The Argumentative Turn Revisited: Public Policy as Communicative Practice. Durham, London: Duke University Press, pp. 305–342.

Jänicke, Martin (1984): Umweltpolitische Prävention als ökologische Modernisierung und Strukturpolitik. Edited by Wissenschaftszentrum Berlin (WZB). Berlin (IIUG Discussion Paper).

Jänicke, Martin (2000a): Chancen und Grenzen von ökologischer Modernisierung. Karriere eines Konzepts. In *Ökologisches Wirtschaften* (6).

Jänicke, Martin (2000b): Ökologische Modernisierung als Innovation und Diffusion in Politik und Technik: Möglichkeiten und Grenzen eines Konzepts. FFU-report 00–01. Forschungsstelle für Umweltpolitik, Freie Universität Berlin.

Johnston, Hank (1995): A Methodology for Frame Analysis: From Discourse to Cognitive Schemata. In Hank Johnston, Bert Klandermans (Eds.): Social Movements and Culture. Minneapolis: University of Minnesota Press (Social movements, protest, and contention), pp. 217–246.

Johnston, Hank (2002): Verification and Proof in Frame and Discourse Analysis. In Bert Klandermans, Suzanne Staggenborg (Eds.): Methods of Social Movement Research. Minneapolis, London: University of Minnesota Press (Volume 16), pp. 62–91.

Johnston, Hank; Klandermans, Bert (1995): The Cultural Analysis of Social Movements. In Hank Johnston, Bert Klandermans (Eds.): Social Movements and Culture. Minneapolis: University of Minnesota Press (Social movements, protest, and contention), pp. 3–24.

Kern, Florian (2011): Ideas, Institutions, and Interests. Explaining policy divergence in Fostering 'System Innovations' Towards Sustainability. In *Environment and Planning C Government Policy* 29 (6), pp. 1116–1134. DOI: 10.1068/c1142.

Kitschelt, Herbert P. (1986): Political Opportunity Structures and Political Protest: Anti-Nuclear Movements in Four Democracies. In *British Journal of Political Science* 16 (1), pp. 57–85.

Koopmans, Ruud (2004): Movements and Media. Selection Processes and Evolutionary Dynamics in the Public Sphere. In *Theory and Society* 33 (3/4), pp. 367–391. DOI: 10.1023/B:RYSO.0000038603.34963.de.

Koopmans, Ruud; Kriesi, Hanspeter (1995): Institutional Structures and Prevailing Strategies. In Hanspeter Kriesi, Ruud Koopmans, Jan Willem Duyvendak, Marco Giugni (Eds.): New Social Movements in Western Europe. A Comparative Analysis. Minneapolis, MN: University of Minnesota Press (Social movements, protest, and contention, v. 5), pp. 26–52.

Koopmans, Ruud; Muis, Jasper (2009): The Rise of Right-Wing Populist Pim Fortuyn in the Netherlands. A Discursive Opportunity Approach. In *European Journal of Political Research* 48 (5), pp. 642–664. DOI: 10.1111/j.1475-6765.2009.00846.x.

Koopmans, Ruud; Olzak, Susan (2004): Discursive Opportunities and the Evolution of Right-Wing Violence in Germany. In *American Journal of Sociology* 110 (1), pp. 198–230. DOI: 10.1086/386271.

Koopmans, Ruud; Statham, Paul (1999): Ethnic and Civic Conceptions of Nationhood and the Differential Success of the Extreme Right in Germany and Italy. In Marco

Giugni (Ed.): How Social Movements Matter. Minneapolis, MN: Univ. of Minnesota Press (Social movements, protest, and contention, 10), 225–252.

Krause, Florentin; Bossel, Hartmut; Müller-Reißmann, Karl-Friedrich (1980): Energie-Wende. Wachstum und Wohlstand ohne Erdöl und Uran; ein Alternativ-Bericht des Öko-Instituts, Freiburg. Frankfurt am Main: Fischer.

Landwehr, Claudia; Böhm, Katharina (2014): Strategic Institutional Design: Two Case Studies of Non-Majoritarian Agencies in Health Care Priority-Setting. In *Government and Opposition*, pp. 1–29. DOI: 10.1017/gov.2014.37.

Leipold, Sina; Feindt, Peter H.; Winkel, Georg; Keller, Reiner (2019): Discourse analysis of environmental policy revisited: traditions, trends, perspectives. In *Journal of Environmental Policy & Planning* 21 (5), pp. 445–463. DOI: 10.1080/1523908X.2019.1660462.

Lockwood, M.; Kuzemko, C.; Mitchell, C.; Hoggett, R. (2016): Historical Institutionalism and the Politics of Sustainable Energy Transitions. A Research Agenda. In *Environment and Planning C: Government and Policy*. DOI: 10.1177/0263774X16660561.

Lukes, Steven (1974): Power. A Radical View. 1. publ. London: Macmillan (Studies in sociology).

March, James G.; Olsen, Johan P. (1984): The New Institutionalism. Organizational Factors in Political Life. In *American Political Science Review* 78 (3), pp. 734–749. DOI: 10.2307/1961840.

March, James G.; Olsen, Johan P. (2008): Elaborating the "New Institutionalism" – Oxford Handbooks. In Sarah A. Binder, R.A.W. Rhodes, Bert A. Rockman, Roderick A. W. Rhodes (Eds.): The Oxford Handbook of Political Institutions. Oxford: Oxford University Press (The Oxford handbooks on political science).

McAdam, Doug (1996): Conceptual Origins, Current problems, Future directions. In Doug McAdam, John D. McCarthy, Mayer N. Zald (Eds.): Comparative Perspectives on Social Movements: Political Opportunities, Mobilizing Structures, and Cultural Framings. Cambridge: Cambridge University Press, pp. 23–40.

McAdam, Doug; Schaffer Boudet, Hilary (2012): Putting Social Movements in their Place. Explaining Opposition to Energy Projects in the United States, 2000–2005. Cambridge, New York: Cambridge University Press (Cambridge studies in contentious politics).

Meyer, David S.; Minkoff, Debra C. (2004): Conceptualizing Political Opportunity. In *Social Forces* 82 (4), pp. 1457–1492. Available online at http://sf.oxfordjournals.org/content/82/4/1457.full.pdf#page=1&view=FitH, checked on 11/2/2016.

Mol, Arthur; Spaargaren, Gert (2000): Ecological modernisation theory in debate: A review. In *Environmental Politics* 9 (1), pp. 17–49.

Morris, Graig; Jungjohann, Arne (2016): Energy democracy. Germany's Energiewende to Renewables. London: Palgrave Macmillan.

Motta, Renata (2015): Transnational Discursive Opportunities and Social Movement Risk Frames Opposing GMOs. In *Social Movement Studies* 14 (5), pp. 576–595. DOI: 10.1080/14742837.2014.947253.

Noakes, John A.; Johnston, Hank (2005): Frames of Protest: A Road Map to a Perspective. In Hank Johnston (Ed.): Frames of Protest. Social Movements and the Framing Perspective. Lanham, MD: Rowman & Littlefield, pp. 1–32.

Partzsch, Lena (2015): Kein Wandel ohne Macht – Nachhaltigkeitsforschung braucht ein mehrdimensionales Machtverständnis. In *GAIA – Ecological Perspectives for Science and Society* 24 (1), pp. 48–56. DOI: 10.14512/gaia.24.1.10.

Rein, Martin; Schön, Donald (1993): Reframing policy discourse. In Frank Fischer, John Forester (Eds.): The Argumentative Turn in Policy Analysis and Planning. Durham, NC: Duke University Press, pp. 145–166.

Rhodes, R. A. W. (2008): Old Institutionalisms – Oxford Handbooks. In Sarah A. Binder, R.A.W. Rhodes, Bert A. Rockman, Roderick A. W. Rhodes (Eds.): The Oxford Handbook of Political Institutions. Oxford: Oxford University Press (The Oxford handbooks on political science).

Schattschneider, Elmer Eric (1960): The Semisovereign People: A realist's view of democracy in America. Hinsdale, IL: Dryden Press.

Schmidt, Vivien A. (2008): Discursive Institutionalism: The Explanatory Power of Ideas and Discourse. In *Annual Review of Political Science* 11 (1), pp. 303–326. DOI: 10.1146/annurev.polisci.11.060606.135342.

Schmidt, Vivien A. (2011): Reconciling Ideas and Institutions Through Discursive Institutionalism. In Daniel Béland, Robert Henry Cox (Eds.): Ideas and Politics in Social Science Research. Oxford, New York: Oxford University Press, pp. 47–64.

Schmidt, Vivien A. (2012): Discursive Institutionalism: Scope, Dynamics, and Philosophical Underpinnings. In F. Fischer, H. Gottweis (Eds.): The Argumentative Turn Revisited: Public Policy as Communicative Practice. Durham, NC, London: Duke University Press, pp. 85–113.

Schmidt, Vivien A. (2015): Discursive Institutionalism: Understanding Policy in Context. In Frank Fischer, Douglas Torgerson, Anna Durnová, Michael Orsini (Eds.): Handbook of Critical Policy Studies. Cheltenham, Northampton, MA: Edward Elgar Publishing (Handbooks of research on public policy), pp. 171–189.

Schmidt, Vivien A. (2017): Britain-Out and Trump-In. A Discursive Institutionalist Analysis of the British Referendum on the EU and the US Presidential Election. In *Review of International Political Economy* 24 (2), pp. 248–269. DOI: 10.1080/09692290. 2017.1304974.

Snow, David A. (2004): Framing Processes, Ideology, and Discursive Fields. In David A. Snow (Ed.): The Blackwell Companion to Social Movements. Malden, MA: Blackwell (Blackwell companions to sociology), pp. 380–412, checked on 1/30/2017.

Snow, David A.; Benford, Robert D. (1992): Master Frames and Cycles of Protest. In Aldon D. Morris (Ed.): Frontiers in social movement theory. New Haven, CT: Yale University Press, pp. 135–155.

Szarka, Joseph (2004): Wind Power, Discourse Coalitions and Climate Change: Breaking the Stalemate? In *European Environment* 14 (6), pp. 317–330.

Tarrow, Sidney (1983): Struggling to Reform. Social Movements and Policy Change during Cycles of Protest: Center for International Studies, Cornell University (Occasional Paper No. 15).

Tilly, Charles (1999): Conclusion. From Interactions to Outcomes in Social Movements. In Marco Giugni (Ed.): How Social Movements Matter. Minneapolis, MN: Univ. of Minnesota Press (Social movements, protest, and contention, 10), pp. 253–270.

Toke, Dave (2002): Wind Power in UK and Denmark: Can Rational Choice Help Explain Different Outcomes? In *Environmental Politics* 11 (4), pp. 83–100. DOI: 10.1080/ 714000647.

Verbruggen, Aviel; Lauber, Volkmar (2012): Assessing the Performance of Renewable Electricity Support Instruments. In *Energy Policy* 45, pp. 635–644. DOI: 10.1016/j. enpol.2012.03.014.

Vrablikova, Katerina (2014): How Context Matters? Mobilization, Political Opportunity Structures, and Nonelectoral Political Participation in Old and New Democracies. In *Comparative Political Studies* 47 (2), pp. 203–229. DOI: 10.1177/0010414013488538.

Walker, Gordon (2011): The Role for 'Community' in Carbon Governance. In *WIREs Climate Change* 2 (5), pp. 777–782. DOI: 10.1002/wcc.137.

Walker, Gordon; Devine-Wright, Patrick (2008): Community Renewable Energy. What Should it Mean? In *Energy Policy* 36 (2), pp. 497–500. DOI: 10.1016/j. enpol.2007.10.019.

WBGU (2011): World in Transition. A Social Contract for Sustainability. German Advisory Council on Global Change (WBGU). Available online at www.wbgu.de/fileadmin/user_upload/wbgu.de/templates/dateien/veroeffentlichungen/hauptgutachten/ jg2011/wbgu_jg2011_kurz_en.pdf, checked on 6/5/2017.

Wolsink, Maarten (2007): Planning of Renewables Schemes: Deliberative and Fair Decision-Making on Landscape Issues Instead of Reproachful Accusations of Non-Cooperation. In *Energy Policy* 35 (5), pp. 2692–2704. DOI: 10.1016/j.enpol.2006.12.002.

Zald, Mayer N. (1996): Culture, Ideology, and Strategic Framing. In Doug McAdam, John D. McCarthy, Mayer N. Zald (Eds.): Comparative Perspectives on Social Movements: Political Opportunities, Mobilizing Structures, and Cultural Framings. Cambridge: Cambridge University Press, pp. 261–274.

3 Renewable energy policy and politics in Canada and Germany

The history of the industrialized world is intrinsically linked to the use of fossil fuels. They served as the basis for the transformation from agrarian to industrialized societies. Blessed with the seemingly inexhaustible energy sources of oil, coal and gas, economies could thrive around their main organizational principle of economic growth. Yet, the reliance on fossil fuels has come at a price. The fossil fuel-based economic pathway has severely contributed to anthropogenic climate change. Replacing fossil-based generation systems by renewable energy sources such as wind power is considered a comparably cost-efficient option to mitigate climate change. Yet, good wind energy resources alone do not automatically turn a jurisdiction into a leader in wind energy deployment. A convenient climate for wind power development include, amongst others, a favorable policy environment including deployment targets and a corresponding support scheme, a supportive administration and a positive societal attitude towards renewable energy.

North America and Western Europe are the two world regions that have contributed the largest share of total cumulative greenhouse gas emissions (Marcotullio et al. 2018). As their energy system still rely to an important degree on fossil fuels and due to their historical contribution, these two regions have a particular responsibility to change the way they generate and use energy. In light of the need to substantially decrease greenhouse gas emissions worldwide, they can serve as examples for other jurisdictions. Germany and Canada are two examples of countries in western Europe and North America where wind energy has flourished under a convenient climate for renewable energy. Both belong to the seven largest economies of the world, built up their economic sectors mainly upon the use of fossil fuels and are now making efforts to decarbonize their energy systems and economies.

This chapter describes Germany and Canada's climate change and renewable energy policy. Germany is one of the countries where renewable energy and wind turbines developed first, and Canada is catching up with diverting its energy systems. Both are federal states. Canadian provinces enjoy sovereignty over energy issues including the support of wind power, while Germany's federal states steer wind turbine development via spatial planning. The chapter discusses in detail the two sub-national forerunner jurisdictions with regard to wind power development, i.e. Brandenburg in Germany and Ontario in Canada. This includes a characterization of their discursive energy space. Brandenburg scores second in Germany and Ontario first in Canada with regard to installed wind power capacity. Ontario's landmass is more than 30 times as large as Brandenburg's, but given the fact that most of the wind turbines are developed in Southwestern Ontario, the effective land surface for wind turbine development is comparable to Brandenburg's. Population density in Southwestern Ontario is, with 86 inhabitants per square km, similar to Brandenburg with 85 inhabitants. The development of wind turbines has benefitted from the same policy scheme of a feed-in tariff. The German federal scheme has served as an example for Ontario's policy makers.

Despite these similarities, Brandenburg and Ontario had different economic starting conditions for developing their wind energy policies. Ontario's GDP is nearly ten times as big as Brandenburg's. While Brandenburg's lignite sector is an important part of its economy and is planned to be phased out only by 2038, at the latest, Ontario has already phased-out its coal plants. The fact that Brandenburg develops renewable energy alongside its lignite sector makes it a typical case for the German energy transition. Ontario has the largest installed nuclear capacity of Canada, while Germany is planning to phase out nuclear power in the coming years. These different takes on energy system change have a crucial impact on the discursive meaning contexts in the two jurisdictions, which will be discussed in a later chapter of the book.

Germany: from forerunner to laggard?

The Federal Republic of Germany is known internationally as a pioneer in fostering renewable energy generation (Morris, Jungjohann 2017). Due to a favorable policy environment, the share of renewable electricity generation in 2017 was 33.3 percent, being followed by lignite (22.5 percent), hard coal (14.1 percent), natural gas (13.2 percent), nuclear (11.7 percent) and others (BMWi 2019). Wind energy contributed 16.3 percent and installed wind power capacity in 2018 was 52,500 MW onshore and 6,400 MW offshore (BMWi 2019). By 2030, the government aims at increasing the renewables share in electricity consumption to 65 percent and to 80 percent by 2050. This means more than doubling the current rate of 37.8 percent renewables in electricity consumption in 2018 (Agora Energiewende 2019). While energy falls under federal leadership, Germany's energy policy is also embedded in the policy framework of the European Union (European Commission 2019b). Germany has been described as a "cognitive leader" in EU renewable energy policy because its domestic policy idea of a feed-in tariff has prompted other countries to adopt the same policy (Solorio et al. 2014), including Austria, the Czech Republic, France, Greece, Spain and Switzerland (Busch and Jörgens 2012). Germany's support of renewable energy is, however, the result of domestic processes rather than the consequence of EU legislation.

Germany's[1] engagement in renewable energy can be traced back to the 1970s when the oil crisis sparked a sudden rise in coal and nuclear generation. The high public salience of environmental issues contributed to the formation of a strong anti-nuclear movement opposing the rapid construction of nuclear and coal-fired power plants. The term *Energiewende* can be traced back to this time. It involves a nuclear phase-out in the short term and coal phase-out in the mid-term, the development and integration of renewable energies, targets for greenhouse gas reduction, and the improvement of energy efficiency (Hake et al. 2015). The term was first coined by the German think tank Öko-Institut in a 1980 report called "*Energie-Wende: Wachstum und Wohlstand ohne Erdöl und Uran*" (Krause et al. 1980).[2]

Phasing out nuclear power had first been decided in 2000 but in 2010 the operating lifetime of nuclear reactors was extended. As a reaction to the 2011 nuclear meltdown in Fukushima, Japan, the government reversed this 2010 decision, agreed upon a complete nuclear phase-out for the year 2022 and decommissioned eight of the 17 reactors immediately. Agreeing on a date for a coal phase-out, by contrast, was a long process. The government installed a commission made up of researchers, politicians and representatives of organized interests including labor, industry and environment. In early 2019, the commission proposed a coal phase-out by 2038, at the latest. Germany's clinging to its coal is one of the reasons why emissions have not materially dropped. As coal represents an important source of its greenhouse gas emissions, a fast coal phase out is mandatory if Germany wants to meet its 2030 emission reduction pledges (SRU 2017).

In late 2019, the Federal Climate Change Act was passed, which endorsed Germany's climate goal to achieve a 55 percent reduction in greenhouse gas emissions by 2030, as compared to 1990 levels. The act prescribes annual reduction targets, allotting annual emission budgets to six economic sectors and introducing an emission trading system for the transport and building sector. This scheme complements the existing emission trading system at the EU level which covers the energy intensive power stations and manufacturing plants that together account for almost half of the EU's greenhouse gas emissions.

However, Germany is not on track to meet its 2030 emission reduction targets (Climate Action Tracker 2019b). Germany's targets are first of all insufficient to meet a fair share in emission reductions according to the Paris Agreement. Second, the implementation of these targets is also lagging behind. Several recent policy choices have led to the concern that Germany may fall back and lose its status as a forerunner in energy system transformation (Kemfert 2017). Most notably, 2018 saw a substantial decline in new installed wind power capacity: the number of new wind turbines and installed capacity halved compared to 2017 (Deutsche Windguard 2019). In 2019, the numbers plummeted further. This development has mainly been attributed to a decline in potential sites and long approval procedures, as many proposed projects have been appealed before court, also by a growing number of opposing citizens' groups (Wehrmann 2019). This stalled development is seriously affecting the target of 65 percent renewable electricity consumption by 2050. Achieving the target would require adding an annual capacity of 4,000 MW wind power and 5,000 MW solar (Agora Energiewende 2018). These numbers mean a doubling of annual addition of solar capacity, but only mean slightly less than the annual addition of wind capacity that took place before the sharp decrease in 2018. Regardless of plummeting wind turbine development, the German government tabled a draft federal law at the end of 2019 to install a general distance of 1,000 meters from wind turbines to the next settlement. This would seriously restrict the possible sites for wind turbines and was meant as a response to a growing amount of public discontent over local wind turbine development that was also put forward by several German state premiers. Facing sharp criticism on the proposal, the government backed away and postponed the decision to 2020.

Despite this recent backlash in political wind turbine support, general approval for developing renewables remains high in Germany. An October 2019 opinion poll revealed that 89 percent of Germans support the further development of renewable energies. Sixty-four percent approve of the construction of renewable energy facilities in the neighborhood in general, while support is higher for solar (66 percent) and lower for wind energy (51 percent) and biogas (33 percent) (AEE 2019b). An earlier poll found that 86 percent of the German population wish for an accelerated or at least constant deployment rate of renewables, but 87 percent wish for consistent or improved possibilities for participation in the *Energiewende* (TNS Emnid 2016). As for wind power, another survey showed that 81 percent consider on-shore wind energy as "important" or "very important" for the energy transition in Germany, while five percent say that it is not important (FA Wind 2016). The survey further showed that 71 percent favor increased public participation, but the survey also revealed a substantial divide between the eastern and western parts of Germany. Only five percent of respondents in the east consider the benefits of on-shore wind for the regional economy as relatively high, opposed to 29 percent in the west (FA Wind 2016). This may be explained by the lower level of community energy projects in eastern Germany as compared to western Germany. In order to understand the way in which the *Energiewende* is anchored in German society and is part of Germany's discursive energy space, it is important to consider the history of wind power in Germany.

The federal feed-in tariff as a central lever

The mid-1970s to 1986 were a pioneering phase of wind power in Germany (Ohlhorst 2009). Small-scale developers contributed significantly to the development of 10 to 50 kW wind turbines, while the government launched its first research and development programs (Ohlhorst 2009). The anti-nuclear movement in Germany saw unprecedented growth after the 1986 Chernobyl nuclear accident, which made the risks of nuclear power tangible for large parts of the German population, at least in West Germany. For instance, the government issued warnings in 1986 not to consume vegetables grown outdoors due to contaminated rainclouds transporting hazardous radiation to Germany. As a political consequence of the catastrophe, the German government halted the commissioning of new reactors and established, six weeks after the accident, the Federal

Ministry for the Environment, Nature Protection and Nuclear Safety. With the establishment of the IPCC in 1988, climate change also reached public awareness in Germany and the government started to regard wind power as a possible way to reduce greenhouse gas emissions. In 1989, the government launched a program to reach 100 MW of installed capacity of wind power within five years. Due to the high demand, the program was extended to 250 MW in 1991 (Ohlhorst 2009). In 1990, there were 300 installed wind turbines in Germany and ten years later, more than 8000 (Ohlhorst 2009).

A major milestone in renewable energy development in Germany was the Feed-In Law ("*Stromeinspeisegesetz*"), which entered into force in December 1990. It represented the first large-scale feed-in tariff (FIT) that considerably promoted renewable energy generation. The act obligated the German power industry to feed renewable electricity into their grids at a 90 percent rate of the average price per kWh derived from power sales. This act was a crucial policy for renewable energy that paved the way for a considerable change of the German power sector. As the incumbent industries possibly did not foresee the policy's major impact on their existing business model, the successful enactment of the act has therefore also been termed a "political accident" (Lauber 2012, p. 47). Indeed, the introduction of the FIT saw a new type of actor entering the arena. While the large incumbent power industries largely remained inactive in developing renewable energy, there were already 221 citizen's wind farms ("*Bürgerwindparks*") in 1989, 1149 two years later and 3625 six years later (Byzio et al. 2005, cited in Ohlhorst, p. 139).

The 1997 change of the Federal Building Code was another milestone in wind power legislation, which now classified wind turbines as "privileged facility" (§35 Building Code). Unless a communal zoning plan or a regional planning program demarcates permissible or non-permissible places for wind turbine development, the erection of wind turbines is generally allowed if those do not lead to conflicts with the public interest. From 1998 to 2002, multiple developments at a national and European level contributed to the booming phase of wind energy (Ohlhorst 2009). In the domestic realm, the Renewable Energy Sources Act (EEG) was most important. The EEG entered into force in 2000 with the aims of further expanding renewable energy deployment, rendering Germany's power supply system more environmentally sound and climate-friendly and decreasing dependence on fossil fuel sources (Federal Ministry for Economic Affairs and Energy 2015). At the core of the EEG was a fixed feed-in tariff for renewable energy, a 20 years purchase guarantee and priority access to the grid. The tariff implied an annual reduction to reflect technological innovation and the feed-in rates for wind varied according to the quality of the wind resource. The EEG considerably ramped up renewable energy production and led to a wind power capacity of 3.2 GW in 2002 (Lauber 2012, p. 50). This capacity was again mainly pushed forward by farmers and energy cooperatives.

The EEG amendments of 2004, 2009 and 2012 took place in a continuing climate of enthusiasm for renewables. Due to the increasing number of wind turbines, issues of social acceptance became evident in some regions. Furthermore, concerns over rising electricity costs started to emerge. Those were associated with the rapid expansion of renewables leading to a rise in the surcharge on electricity prices imposed under the EEG. In practice however, they were also caused by declining wholesale power prices, which increased the level of support needed to cover the total generation costs of renewables. As a consequence of the discussion on electricity costs, the 2017 EEG amendment brought about a fundamental change to the renewable energy policy framework in Germany. It replaced the feed-in tariff for wind energy by an auction system, which started in January 2017.

This substantial change has been put into context with "effective steering" and "restrictions on costs" (Federal Ministry for Economic Affairs and Energy 2016). The discussion on this major change to the EEG was highly polarized. Illustrative is the coverage of two major nationwide daily newspapers. The center-left daily newspaper *Süddeutsche Zeitung* headlined that "Germany is Putting its Energiewende in Chains" *("Deutschland legt seiner Energiewende Fesseln an")* (SZ 2016), which refers to the concern that the reform overly restricts Germany's ambitions for energy system change. By contrast, the conservative daily newspaper *Frankfurter Allgemeine Zeitung* titled "Billions into the Wind" *("Milliarden in den Wind")*, referring to the discussion over the high costs of Germany's energy transition (FAZ 2016). These two headlines show that some regarded the EEG 2017 as an unnecessary slow-down of the energy transition in Germany, while others applauded the change as a measure to limit costs.

The role of the federal states

While the EEG at the national level has served as a major impetus for renewable energy development, Germany's 16 federal states *("Länder")* have substantial leverage in steering renewable energy development. First of all, the states take part in the decision-making via the regular legislative procedure including deliberation in the *Bundesrat* as the central organ of the *Länder* representation and Germany's second parliamentary chamber. In the 2016 negotiations on the EEG amendment, the Bavarian state premier Horst Seehofer for example successfully demanded further inclusion of bioenergy in the portfolio of supported renewable energy sources. Beyond this legislative deliberation, the states also have considerable leverage in steering renewable energy development by regional or municipal spatial planning. After 2002 for example, national onshore wind expansion rates decreased, mainly due to restricting planning initiatives by some federal states that saw saturation in their jurisdiction in relation to the landscape view, which was also regarded as limiting social acceptability (Ohlhorst 2009).

A major attempt by German federal states to address declining levels of social acceptance of wind power occurred in 2014. The two federal states of Bavaria and Saxony successfully started a *Bundesrat* initiative to tilt the federal Building Code towards a greater autonomy for federal states to determine standardized setbacks to wind turbines, i.e. the distance a wind turbine must be placed from housing or other land uses. Both federal states wanted to impose a general setback of ten times the height of a turbine to the next residential area ("10h" rule) in their jurisdiction. This generally counteracts provisions in the Federal Building Code, which requires setbacks to be determined by regional or municipal planning. The opportunity for federal states to implement standardized setbacks known as "*Länderöffnungsklausel*"[3] ("Federal State Opening Provision") was granted temporarily from August 2015 until the end of December 2015. Bavaria was the only federal state that ultimately implemented this "10h" setback and as a result, it has nearly halted further wind energy development in its jurisdiction (Hehn and Miosga 2015).

Invoking this temporary possibility of introducing a standardized setback was one response by federal states to declining levels of local acceptance to wind in some regions. Other federal states have acted differently. Mecklenburg-West Pomerania has for example passed the "*Bürger- und Gemeindenbeteiligungsgesetz*" ("Participation Act for Citizens and Municipalities") in 2016. The act requires each wind developer to offer at least 20 percent of its project shares to local residents within a 5 km radius of a wind turbine at a maximum rate of 500 Euro (Mecklenburg Vorpommern 2016). Similarly, the Brandenburg state government passed, in June 2019, the *Windenergieanlagenabgabengesetz* (BbgWindAbgG). This "Wind Turbine Levy Act" obliges the operator of each wind turbine becoming operational from 1st of January 2020 to pay an annual levy of 10,000 Euro to the surrounding municipalities (Landtag Brandenburg 2019). Another example is the federal state of Thuringia, which has introduced a voluntary certificate "*Faire Windenergie*" ("fair wind energy"). The public energy agency awards this certificate to specific wind projects to signal a good level of communication, consultation and participation of local residents and communities (Thüringer Ministerium für Umwelt, Energie und Naturschutz 2016).

Brandenburg: a coalition of coal and wind power

The federal state of Brandenburg is situated in the north-eastern part of Germany, enclosing the German capital city of Berlin. Brandenburg is one of the five new German federal states that were re-established after the reunification of the German Democratic Republic (GDR) with West Germany in 1990. Brandenburg stretches across 29.479 km^2 and represents with a total population of 2.5 million and 84.9 inhabitants per km^2 the second least populated federal state in Germany (European Commission 2019a, end of 2018). The federal state is largely rural with vast tracts of open landscapes dominated by agriculture and nature protection areas. Next to Brandenburg's capital city of Potsdam, Brandenburg only has a few medium-sized urban centers and smaller villages characterized by a compact settlement structure. With a gross domestic product (GDP) of 69.4 billion euro in 2017, Brandenburg has a structurally weak economy that represents 2 percent of the national total (European Commission 2019a). Brandenburg's unemployment rate of 4.1 percent (in 2018) is comparable to the national average of 3.4 percent and well below the European average of

6.9 percent (European Commission 2019a). Poor economic perspectives are most evidently visible in Brandenburg's remote rural outer parts, which suffer from population decline, while its inner parts close to Berlin have benefitted from the capital city's recent population growth.

The social democratic SPD (*Sozialdemokratische Partei Deutschlands*) has been the strongest political force in Brandenburg since reunification and has since been the party of Brandenburg's prime minister. Brandenburg's prime ministers relevant for this study are Matthias Platzeck (who served from 2002 to 2013) and Dietmar Woidke (in office since 2013). Especially the former has been an important driving force behind renewable energy development in Brandenburg. The SPD governed in coalition with the conservative CDU (*Christlich Demokratische Union Deutschlands*) from 1999 until 2009, with the left-wing party DIE LINKE from 2009 to 2019 and with CDU and the center-left Green Party (*Bündnis 90/die Grünen)* since the recent elections of 2019. Next to SPD (25 seats in state parliament), CDU (15 seats) and Green Party (10 seats), three other parties are represented in state parliament. Those include DIE LINKE (10 seats), the center-liberal BVB/Freie Wähler (five seats), and the right-wing AfD (*Alternative für Deutschland*, 23 seats). The latter was first elected into state parliament in 2014 and has recently gained leverage at the federal level and in other federal states in Germany as a populist response to the rising numbers of refugees to Germany. The party is the only political force in Brandenburg state parliament which openly stands for the abolition of the federal Germany's Renewable Energy Sources Act. Climate change was not mentioned in its 2014 electoral program and its 2019 electoral program denied anthropogenic climate change (AfD Brandenburg 2014, 2019).

Brandenburg's energy sector has historically been heavily dominated by lignite mining and lignite-fired electricity generation. The southern region of Lusatia, which became part of Brandenburg upon reunification, served as a crucial energy provider during GDR times. Lignite mining and combustion still represents a major pillar of Brandenburg's energy mix. In 2017, the largest share of Brandenburg's gross electricity production consisted of 57.3 lignite and 32.4 renewable energy sources (AEE 2019a). Lignite mining and combustion represents a substantial component of Brandenburg's economy, contributing almost 40 million euros in tax revenue for the state and municipal level and providing more than 10,000 direct and indirect jobs, especially in the region of Lusatia (MWE 2012, p. 44). These numbers have continually declined and are still expected to drop further. Due to its economic importance for the state of Brandenburg and the employment opportunities it offers in a region that would otherwise suffer from economic decline, the state government has long been reluctant in setting a date for a coal phase-out. This reliance on fossil energy sources comes with high energy sector-related CO_2 emissions. As of March 2016, Brandenburg's energy sector emitted 58.0 million tons of CO_2 which made Brandenburg's per capita CO_2 emissions more than twice the national German average (LUGV 2016). In 2014, Brandenburg accounted for 4.8 percent of the total German primary energy consumption, but was responsible for 7.8 percent of energy-related CO_2

emissions (MWE 2016, p. 28). The fact that Brandenburg exports more than half of its generated power (ZAB 2014, p. 86) is a contributing factor.

Despite this historic attachment to lignite, Brandenburg has become a national forerunner in wind power development. This makes Brandenburg an emblematic case for the characteristics of the German energy transition, as the state government has long supported fossil fuels and renewables in parallel (see also Bafoil 2016). For its efforts to increase the share of renewable energies, Brandenburg was awarded the bi-annual lodestar award (*"Leitstern"*) on three consecutive occasions from 2008 to 2012, a prize for the German federal state that achieves the best results in supporting and expanding renewable energies. The explanatory statement praised an exemplary political agenda for renewable energies, strong performance in the expansion of wind energy, and strong efforts in the reduction of barriers for renewable energy development (Agentur für Erneuerbare Energien 2012). Due to Brandenburg's high rate in renewable energy generation, the electric vehicle company Tesla proposed, in late 2019, to build a large automobile factory in Brandenburg.

The overall policy framework upon which Brandenburg's renewable energy sector flourished has been the federal Renewable Energy Sources Act (EEG, since 2000) and the previous Feed-In Law (1990–2000). Nonetheless, the government of Brandenburg has become pro-active in supporting the development of renewable energy far beyond this national impetus. Brandenburg started to grant financial support to wind energy soon after the state had been re-created upon German reunification. From 1991 to 2000, Brandenburg supported wind energy with 82 million *Deutsche Mark* (MLUR 2000, p. 8), which is approximately 42 million euro, not accounting for inflation. The first targets for renewable procurement were issued in the context of the need to reduce CO_2 emissions. In August 1994, the communications department of the Brandenburg Ministry for the Environment, Nature Protection and Regional Planning (MUNR) referred to the national targets of reducing CO_2 emissions by 25 to 30 percent until 2005 and concluded that next to biomass, wind power would be the most widely available and cheapest renewable energy source in Brandenburg (MUNR 1994, p. 14). Brandenburg's energy concept announced in 1996 set the target for renewable energy to reach five percent of primary energy consumption by 2010, emphasized energy efficiency and the use of renewable energy in order to secure employment and decrease CO_2 emissions by about 40 percent from 1990 levels (MUNR 1996a, p. 27). Furthermore, the energy concept proclaimed a stabilization of lignite mining on the level of 35–40 million tons per year, while resource conservation and minimizing of environmental impacts "should have top priority" (MUNR 1996a, p. 27). In 1998, the Brandenburg government endorsed the target of five percent renewable energy until 2010 and also launched the REN financial support program *("REN-Programm: Rationelle Energieverwendung und Nutzung erneuerbarer Energiequellen")* (MUNR 1997). The main underlying rationale was to decrease air pollutants including CO_2, to support future-proof technologies (*"zukunftsfähige Technologien"*) and to secure modern employment opportunities (MUNR 1997).

By the end of 1994, Brandenburg already had approximately 100 wind turbines, out of which 60 had received financial support as pilot projects (MUNR 1994, p. 14). Since these early beginnings of wind energy in Brandenburg, both installed capacity and numbers of wind turbines have increased dramatically. While in July 1995, Brandenburg's installed total wind power capacity amounted to 38 MW (MUNR 1995, p. 19), it increased more than tenfold by 2001 (Landesregierung Brandenburg 2002, p. 10). In the year 2000, Brandenburg had 615 wind turbines, and ten years later, 2,957 (Landtag Brandenburg 2014). This sharp increase is also attributable to Brandenburg's ambitious renewable energy targets and programs in the context of its Energy Strategy 2020 and 2030 ("*Energiestrategie 2020*" and "*Energiestrategie 2030*"), which far exceeded the earlier relatively modest targets. In 2018, Brandenburg was home to 3,821 wind turbines and had an installed wind power capacity of 7,081 MW, placing it second after the federal state of Lower Saxony with 11,165 MW and before Schleswig-Holstein with 6,964 MW (Deutsche Windguard 2019).

Brandenburg's energy strategies 2020 and 2030

Brandenburg's energy policy takes the form of "Energy Strategies" that are not legally binding but present the envisaged commitments in the energy sector, including procurement targets for each energy source, CO_2 reduction targets and measures for energy efficiency. While the Energy Strategy 2010 (issued in 2002) still considered the earlier goal of five percent renewable energy in primary energy consumption until 2010 as an "appropriate contribution" (Landesregierung Brandenburg 2002, p. 23) to the national and European strategy to increase the use of renewable energy, the Energy Strategies 2020 and 2030 went far beyond this target. The Energy Strategy 2020 (issued in 2008) identified, for the first time in Brandenburg, renewable energies as one of the "supporting pillars" ("*tragende Säule*") of Brandenburg's energy sector (MW 2008, p. 31). The strategy announced the target of raising the share of renewables in Brandenburg's primary energy consumption to 20 percent and ramping up installed wind power capacity to 7,500 MW by 2020 (MW 2008). The strategy defined a surface area of 555 km^2 as necessary for achieving these procurement targets, which would represent an increase of 50 percent to the then existing wind areas (MW 2008, p. 45). The Energy Strategy 2030 (MWE 2012) extended those targets, in that the share of renewable energies in primary energy consumption should now account for 32 percent by 2030. As for wind energy, installed capacity shall be raised to 10,500 MW by 2030 (MWE 2012, p. 39) for which two percent of the landmass would be necessary. This target meant more than doubling the then existing installed wind power capacity, which amounted to 4,600 MW at the end of 2011 (BWE 2016).

In order to support this target, the Energy Strategy 2030 emphasizes the importance of system integration by grid expansion and the development of energy storage technologies. Supporting corresponding research and development, the strategy announces the launch of the program RENPlus as a continuation of the previous REN program. RENPlus is a financial support program for renewable energy-related measures, such as investments in energy efficiency or the development of energy storage. These endeavors in research and development are part of a plan to intensify cooperation with Berlin. The Energy Strategy 2030 emphasizes the aim to jointly become an internationally recognized forerunner region in regional energy system change and technology. The region of Berlin-Brandenburg is envisioned to become a "region of the energy transition" ("*Region der Energiewende*" (MWE 2012, p. 30), with Brandenburg as forerunner in renewable energy and grid development and Berlin as energy sink. This cooperation shall serve as an innovative model to be transferred to other regions and builds upon Brandenburg and Berlin's existing cooperation in the realms of regional planning, public transport or research on energy technology development.

Both energy strategies spell out reduction targets for greenhouse gas emissions which roughly correlate with the then national targets. They also both attribute an important role for lignite mining alongside the deployment of renewables. Lignite mining and combustion are mainly described as inexpensive and reliable in terms of contributing to the security of power supply. Lignite is assigned the role of contributing to two central targets of the jurisdiction's energy policy which, in the Energy Strategy 2020, is organized as a "target triangle" consisting of "security of supply", "economic efficiency" and "environmental and climate protection" (MW 2008). The Energy Strategy 2030 regards the national decision to phase out Germany's nuclear fleet until 2022 as a further endorsement of Brandenburg's lignite sector. As Brandenburg exports more than 50 percent of its generated electricity and more than 60 percent of its generated refinery products, the strategy emphasizes Brandenburg's importance for the national security of energy supply (MWE 2012).

The Energy Strategy 2030 defines fossil energy sources as "bridge technologies" towards a renewable energy-based energy system: "the transition towards a sustainable energy supply mix will need to be accompanied by conventional technologies (gas, coal etc.) until a secure energy supply at low prices can be guaranteed by renewables" (MWE 2012, p. 42). It is argued that fossil sources should be further pursued until renewables prove to be cost-effective and provide enough base-load capacity. The "length of the bridge" is described as depending on further innovations in and development of renewables.

Rationale for renewables: ecological modernization and an identity of "Energieland"

As the government of Brandenburg has long been reluctant in setting a date for a coal phase-out before it was decided at the federal level, the question emerges why the Brandenburg government has still issued such ambitious programs to support wind energy. This section presents the corresponding storylines behind the political support that wind power has enjoyed. In Brandenburg as elsewhere, successful wind energy deployment has been a product of several factors. A representative of the Ministry for Economic Affairs and Energy hints to the combination of political will, historical aspects and technical feasibility (Anonymous interview ministry representative). Renewable energy had been politically promoted early-on by the former Prime Minister Matthias Platzeck, who had been State Minister of the Environment before. Second, Brandenburg had a historically strong energy sector based on lignite that would not exist indefinitely and finally, Brandenburg had very good wind resources. While the latter aspect is a crucial prerequisite of a jurisdiction's ability to become a fore-runner in wind energy deployment, the focus here is on the reasons behind a political atmosphere that embraced renewable energy development in Brandenburg.

The fact that Brandenburg has championed renewable energy can be explained by the economic importance of Brandenburg's lignite sector and its dim future prospects as one of the major sources of CO_2 emissions in Germany. Economic considerations are thus at the core of Brandenburg's support for renewable energy. In this vein, the 2009–2014 State Minister of Economic and European Affairs Ralf Christoffers refers to Germany's federal Renewable Energy Sources Act (EEG) as primarily an "industrial policy". Well aware that lignite mining would not be continued indefinitely, the promotion of renewable energy would set the foundation for a green economy ultimately replacing coal as the current major economic sector in Brandenburg (Interview, Ralf Christoffers). In the same vein, a staff member of the ministry regards the EEG as a "wonderful instrument for business development", which has led to "great business models" (Anonymous interview, ministry representative). The federal EEG is thus regarded as a major lever to spur green economic development.

The historic and persisting economic importance of Brandenburg's lignite sector has become enshrined in representations of Brandenburg as an "energy state" ("*Energieland*") or "energy export state" ("*Energieexportland*") (Landtag Brandenburg 2006; MWE 2012). Both convey the strong economic importance of Brandenburg's energy sector and are put forward by the state government to justify ambitious energy policies that stress both Brandenburg's fossil fuel iden-tity and the need to increase the rate of renewables. The commitment to remain-ing an "energy export state" is for example postulated in the 2014 coalition agreement between SPD and DIE LINKE:

> Brandenburg is an energy exporting state and a pioneer in the expansion of renewable energies. It should remain that way in the future. That is why the coalition is committed to the energy transition and the move towards renew-able energies.
>
> (SPD Brandenburg and DIE LINKE Brandenburg 2014, p. 17)[4]

This excerpt shows that *because* Brandenburg shall continue to export electri-city, renewable energies shall be endorsed. This line of argumentation also serves as a justification to pursue a political agenda that puts economic object-ives at the core of Brandenburg's renewable energy policy. The anticipated win-win situation between the economy and the deployment of renewables is in fact a central legitimation behind the Energy Strategy 2020, which was the first strategy to introduce ambitious procurement targets. In his introductory note to the strategy, the then State Minister of Economic Affairs Ulrich Junghanns states that "Brandenburg needs economic growth", and that environmentally friendly energy generation is not only essential against the backdrop of global climate change, but also offers substantial opportunities for economic growth. Therefore, Brandenburg commits to climate protection "together with the economy, not against the economy" (MW 2008, p. 3). Energy policy in Brandenburg is "geared towards promoting the creation of jobs and economic prosperity for the citizens of Brandenburg through technological innovations 'Made in Brandenburg'" (MW 2008, p. 3). The Energy Strategy 2030 endorses this view, stating that Brandenburg's energy policy has constantly been adapted to the dynamic development of the energy sector because "as an energy export and energy transit state" the goal is "to secure economic opportunities and preserve jobs in Brandenburg, to secure economic competitiveness and take its responsibility in the context of national security of energy supply and climate change policies" (MWE 2012, p. 9).

In summary, Brandenburg's wind energy program was heavily informed by the discourse *on ecological modernization*. This involves a win-win narrative between economic and environmental objectives and follows a logic of achieving sustainability by restructuring the economic system in a more environmentally friendly direction with a strong role of the state (Dryzek 2013; Jänicke 1984). Brandenburg's policies were the result of a strong governmental initiative to support the development of a new industry while at the same time delivering on European and national commitments to expand the use of renewable energy. The economic argument had been at the core of Brandenburg's policy which can be explained by the economic importance of its lignite industry. Renewable energy is intended to serve as a way to alleviate the associated structural challenges that Brandenburg faces once lignite mining comes to an end. In this regard, the Energy Strategy 2020 and 2030 represent strong governmental initiatives to spur renewable energy in Brandenburg and "ecologically modernize" Brandenburg's economy to potentially compensate for the losses in employment and tax revenues when Brandenburg's coal phase-out would eventually become a reality.

Implementation: Brandenburg's regional planning approach

While Brandenburg's Energy Strategies are embedded in the overall federal energy policy of the EEG, Brandenburg has considerable leverage in steering wind power development through wind turbine planning. Brandenburg's effort to reduce barriers for wind turbine development was also one of the reasons why the state was granted the lodestar award for its achievements in renewable energy development. In order to understand Brandenburg's success in wind power development, it is therefore necessary to consider its planning approach to wind energy and describe its space of participation.

Wind energy planning in Brandenburg is mainly carried out at the regional planning level. Since 1993, Brandenburg's regional planning is in the hands of five regional planning authorities (RPG: *"Regionale Planungsgemeinschaft"*) whose members are the corresponding *Landkreis* and non-associated towns.[5] An RPG's political decision-making body is the regional assembly which usually meets twice a year. It consists of the *Landkreis* commissioners, the mayors of the non-associated towns and the mayors of municipalities with more than 10,000 inhabitants, and a number (up to a maximum of 40) of councilors that are elected by the *Landkreis* councils and the non-associated towns. Upon request, representatives of chambers, associations and other institutions associated with regional development may be admitted as non-voting, advisory regional councilors to the regional assembly. In the RPG of Uckermark-Barnim for example, representatives of two citizens' initiatives as well as the regional wind power association are advisory members.

Wind power planning in Brandenburg follows the heavily prescribed procedure of designating wind suitability areas (*"Windeignungsgebiet"*) outside of which wind turbine development is foreclosed. Put shortly, the office of each RPG (*"Regionalplanungsstelle"*) proposes a draft of the regional plan containing the precise number and location of wind suitability areas and presents it for approval to the regional assembly. After discussion, revision and adoption by the regional assembly, the regional plan is submitted for approval to the state planning authority, which validates whether the plan is in line with the overarching state development plan. Relying on the local input of the regional assemblies, the planning process of wind energy in Brandenburg has in principle the face of an inclusive and local process.

Nonetheless, the procedure is heavily determined by and embedded in state guidance, national law and court rulings which significantly restrict the autonomy of the RPGs and their regional assemblies. The requirements for wind power planning in Brandenburg have contributed to the fact that the development of regional plans has taken more than ten years in some of Brandenburg's RPGs. By issuing directives, the state of Brandenburg has considerably shaped the way in which the five RPGs carry out wind energy planning. In particular, the 1996, 2009 and 2011 directives were geared towards supporting the designation of wind suitability areas and fostering wind energy development in Brandenburg. The 1996 Wind Energy Directive (*"Windkrafterlaß"*) for example aimed at facilitating and streamlining the wind turbine planning process by introducing the system of wind suitability areas in the first place (MUNR 1996b).[6] Wind turbines were determined to be generally permissible within a wind suitability area (*"Eignungsgebiet"*), but might under certain conditions also be built in restriction areas (*"Restriktionsbereich"*), but never in taboo areas (*"Tabubereich"*). This change replaced a system in which each proposed wind energy project needed to be assessed individually to determine whether it was permissible within existing nature protection and state planning legislation. With this new streamlined approach to wind turbine approvals, Brandenburg aimed at endorsing a wind turbine-friendly politics without compromising nature protection and landscape conservation (MUNR 1996c, p. 9). This "clearance of obstacles" in "unproblematic areas" was regarded as compensation for declining financial funding for wind turbine development due to tight public budgets (MUNR 1996c, p. 9).

Two other important state directives that substantially shaped wind energy planning in Brandenburg were the 2009 wind energy directive (*"Windkrafter-lass"*) and the 2011 directive on setbacks to species. In order to implement the wind energy procurement targets specified in the Energy Strategy 2020, the 2009 directive instructs the RPGs to demarcate approximately 1.9 percent of their landmass as a wind suitability area, but also recommends a setback of 1,000 meters to the next residential area (MIR and MLUV 2009). Unless the submitted regional plans do not designate about 1.9 percent of their landmass as wind suitability area, the overarching state development office no longer approves the RPG's regional plan. This has increased the pressure on the RPGs to identify sufficient areas for wind turbine development. In order to facilitate this process, a 2011 directive introduced standardized setbacks to particular species (*"TAK-Erlass"*) that lowered the previously required distances of wind suitability areas to specific species (MUGV 2011). Setbacks to particular species and habitats are an inherent component of the planning procedure and had previously been decided upon at the single case level. The directive also generally allowed the designation of wind suitability areas in forested areas, including official landscape protection areas (*"Landschaftsschutzgebiet"*) after a thorough assessment of the single area (MUGV 2011). The 2011 directive has sparked a great debate in Brandenburg and garnered criticism from nature protection associations (e.g. NABU Brandenburg and BUND Brandenburg 2010).

Next to those exemplary state directives, various European directives, changes in national laws and court rulings contributed to the need to frequently adapt and correspondingly change Brandenburg's regional planning procedure (Overwien and Groenewald 2015). A mandatory and time-consuming component of the development procedure of regional plans is public consultation, which the Federal Planning Act introduced in 2008. Among the most important changes was also the 1997 amendment in the Building Code, which classified wind turbines as "privileged facility" (§35 Building Code). A series of court rulings by the Federal Administrative Court specified the consequences for wind turbine planning (FA Wind 2015; Overwien and Groenewald 2015) which also apply to Brandenburg. A 2009 ruling prescribed a coherent planning process on the basis of "hard" and "soft" taboo criteria involving three steps (FA Wind 2015). The first encompasses the identification of hard and soft taboo zones. "Hard taboo zones" effectively exclude the development of wind turbines, such as particular nature protection areas or former military areas. In "soft taboo zones", wind turbine development is generally admissible, but regionally specific factors may foreclose it. This can relate to prescriptions made in regional or state development plans with regard to, for example, historically important landscape elements. The second step includes a detailed consideration of the remaining areas by including public concerns or other aspects that may be raised against wind turbines in a specific area. As a final third step, the planning process must result in providing "substantial space" for wind energy in order not to countervail the privileged status of wind turbines guaranteed by the federal Building Code. In 2010, the Higher Administrative Court Berlin-Brandenburg (OVG: "*Oberverwaltungsgericht*") invalidated the wind energy planning carried out in the RPG of Havelland-Fläming by criticizing its deviance from this required planning procedure (OVG Berlin-Brandenburg 2010). As the RPG of Prignitz-Oberhavel had employed the same methodology, they terminated the ongoing planning process and also started a new process.

What is the role of municipalities in Brandenburg's wind turbine planning process? Generally, the municipalities may issue a municipal plan that demarcates land for different purposes. Yet, the strict hierarchy of state development plan over regional plan and municipal plan implies that during the process of devising the regional plan, the RPGs do not necessarily need to follow the municipalities' planning. A representative of the Brandenburg Association for Towns and Municipalities (STGB Brandenburg: "*Städte- und Gemeindebund Brandenburg*") states that the inclusion of the municipalities at this stage of planning is also a matter of which respective regional planner is in charge. In some of the five RPGs, the head regional planner literally tours the municipalities to reach a consensus within the possible planning framework, while in other RPGs, regional planners merely refer to a decision formerly voted on by regional assemblies on whether or not to include a municipality's own planning (Interview representative, STGB). As set out earlier, municipalities with more than 10,000 inhabitants have a seat in the regional assembly. Thus, all municipalities in close vicinity to Berlin have the right to vote on draft versions of the regional plans and the location of wind suitability areas, whereas the rural municipalities, which are host to most of the wind turbines, usually do not, unless they are represented via their *Landkreis* commissioner. Having the status of "affected public body", municipalities are still consulted in the context of the official consultation process, but they do not necessarily have a seat in the regional assembly.

A 2014 survey by the STGB Brandenburg revealed that many mayors were disappointed by the way in which wind turbine development unfolds in Brandenburg. The average commercial tax income per turbine was 1,100 Euro (Kunze 2015) which represents a fraction of their generated revenue. While financial participation of municipalities is not prescribed in the regional planning process, there are several examples of municipalities in Brandenburg that actively engage in renewable and wind energy development in their jurisdiction, although they are few and far between (e.g. Busch and McCormick 2014). One example is the municipality of Schipkau, which developed wind energy on post-mining areas and actively engaged in finding a solution to have the citizens and the municipality participate financially. To this end, a system was set up in which the investor pays 80 Euro per year to each resident for the initial five years. There is no citizen's initiative against wind turbines and as of April 2015, the wind farms in the area of the municipality summed up to 58 wind turbines with a total capacity of 100 MW (Gemeinde Schipkau 2016).

Summary: Brandenburg's discursive energy space

Drawing on the analysis of Germany's and Brandenburg's energy sectors, how can Brandenburg's discursive energy space be characterized? Brandenburg's energy policy is heavily embedded in the federal German policy framework including the FIT and the climate and renewable energy targets. At the same time, Brandenburg has considerable leverage to steer wind turbine development via its regional planning approach and has been applauded for the active role it has taken in supporting wind energy. Brandenburg presides over a sophisticated system of regional planning based upon the designation of wind suitability areas. Within this space of participation, local input is possible, but in practice it is heavily restricted by a number of state directives, national legislation and court rules. Furthermore, the German *Energiewende* discourse is very strong in Brandenburg. It represents an important part of Brandenburg's meaning context and largely follows an *ecological modernization* discourse. It informed the government's renewable energy policies and was further specified by the *Energieland* discourse which played an important role in fostering Brandenburg's role as an energy exporting state. As will be discussed in a later chapter, the characteristics of the discursive energy space left Brandenburg's anti-wind movement in a weak discursive position, unable to effectively scale-up their concerns to the sub-national level.

Canada: towards becoming a forerunner in climate action?

The federal state of Canada is the world's second-largest country by area and is made up of ten provinces and three northern territories. The three territorial governments receive their power as delegated by the federal government, while the division of power between the federal level and the provinces is outlined in the Constitution Act, 1867. While sectors of national concern such as foreign affairs or telecommunications fall under federal jurisdiction, the provinces have sovereignty over sectors such as education, healthcare, and also energy. The generation, transmission and distribution of electricity are matters of provincial legislation. The provincial level is therefore key to understanding energy politics in Canada. Canada's provinces and territories are highly diverse in resource endowment and consequently differ in their energy generation, trade and use. Four provinces and one territory (British Columbia, Manitoba, Newfoundland and Labrador, Quebec, Yukon) derive over 80 percent of their electricity from hydro-electric sources, while Ontario, New Brunswick, Prince Edward Island and Northwest Territories rely on various combinations of power sources. Alberta, Saskatchewan, Nova Scotia, and Nunavut generate the majority of their electricity from fossil fuels, including coal, natural gas, or petroleum (Canada Energy Regulator 2019). Renewable energy sources provided approximately 67 percent of Canada's electricity generation in 2017, with hydro-electric providing 60 percent (Natural Resources Canada 2019b). Wind power generation grew from a negligible amount in 2005 to about four percent of electricity generation, representing one of the fastest growing sources of electricity. Canada is home to two percent of the world's installed capacity of wind power and scores ninth place in the world for wind installations (Germany with ten percent scores third place, after China and United States) (Natural Resources Canada 2019b). As of December 2018, Canada had an installed capacity of 12,816 MW of wind power, with Ontario (5,076 MW) scoring first and Quebec (3,882 MW) second (CanWEA 2019a).

Due to its diverse geography, the whole range of climate change impacts are visible in Canada: Increasing heat waves and droughts throughout the country, floodings and erosion at the coastlines, wildfires in its forested areas, and a thawing permafrost and changing Canadian Arctic in the north. At the same time, Canadian per capita CO_2 emissions are among the highest in the world. The oil and gas sector accounted for more than a quarter (27 percent) of all greenhouse gas emissions in Canada in 2017, closely followed by the transport sector (24 percent) (Environment and Climate Change Canada 2019a). Canada is the fourth largest producer and fourth largest exporter of oil in the world, and Canada's oil sands alone account for 11 percent of Canada's greenhouse gas emissions (Natural Resources Canada 2019a). The high Canadian per capita CO_2 emissions can be attributed to severe winter conditions in large parts of the country and a vast, but unevenly populated country which increases the importance of transportation. Yet, high emissions can also be traced back to a general perception of abundant electricity sources and corresponding consumption patterns. Already in the early 1980s – with reference to the past century – Hooker et al. (1981, p. 23) noted that:

> During this period Canada has also enjoyed a high rate of economic growth, based in part on ready access to cheap energy resources. Canadians have generally acted as though energy would always be available in the forms required, in essentially unlimited quantities, at relatively little expense, and with low social and environmental costs. Such implicit assumptions continue to underlie Canadian social and economic activities. They are evident in car-oriented cities, poorly designed buildings, and supermarket packaging.

While little may have changed since then with regard to the design of cities, buildings or packages, concern for the environment is nevertheless increasing. The latest indication of this is the importance of environmental issues in the 2019 federal election. An October 2019 poll, a few weeks before the election, revealed that climate change was among the top three issues that mattered for voters, a rating that had never been met in previous elections (Shah 2019). Already in 2010 under then Prime Minister Stephen Harper, an opinion poll showed that 66 percent of Canadians were not satisfied with the attention their federal government paid to the environment (Ipsos Reid 2010). The discontent with the environmental performance of Stephen Harper certainly contributed to the 2015 change in government, when Liberal Justin Trudeau won office. In the latest 2019 federal election however, the Liberals lost their majority, forming a minority government. Due to the high public salience of environmental issues during the run-up to the elections, the Green Party won a historic three seats.

Federal climate action . . .

When Premier Justin Trudeau took office in 2015, environmental matters scored high on the agenda. "Canada is back, my good friends", he announced at the 2015 COP-21 climate negotiations in Paris. In Canada's National Statement, he signaled greater leadership in climate action and in the transition towards a low-carbon economy "that is necessary for our collective health, security, and prosperity" (Prime Minister of Canada 2015). Trudeau had only been elected prime minister two months earlier. The October 2015 federal elections ended almost a decade under the conservative government of Stephen Harper (2006–2015), which had gained sad notoriety from a climate and environmental perspective. According to the University of Toronto associate professor Stephen Scharper, Harper's government had "launched an anti-environmental crusade that resembled a scorched-earth campaign" (Scharper 2015, p. 1). The government dismissed more than 700 environmental scientists from Environment Canada, the federal Department of the Environment (Scharper 2015). Additionally, the government stalled and restructured climate science research in Canada by considerably diminishing funding opportunities, reducing opportunities for the media to access scientists and by appointing three climate change skeptics to key granting agencies for university-based scientific research (Cuddy 2010).

Harper's inaction on climate change mitigation peaked in the December 2011 announcement of Canada's official withdrawal from the Kyoto protocol, arguing that the country could not meet its emission reduction goals and had to avoid the fines for non-compliance. Harper argued that such a treaty would not make sense if the biggest emitters US and China were not included and combatting climate change was compromising jobs and economic growth (The Guardian 2011). At the time, Canada ranked among the top ten greenhouse gas emitters and scored first with regard to per capita emissions (Ge et al. 2014, data for 2011). In dishonor of Canada's non-commitment to climate change mitigation, environmentalists awarded the Canadian government the "Lifetime Unachievement Fossil Award" at the 2013 COP-19 climate negotiations in Warsaw.

With the 2015 electoral victory of the Liberal Party, a "climate of hope returns" (Scharper 2015, p. 1), as the new government pledged to make climate change a top priority and called renewable energy a "tremendous opportunity" (CBC 2016). Yet, four years later, it becomes evident that the Liberals were not successful in making climate change enough of a priority. Canada is not on an emission reduction pathway that would meet a fair and sufficient contribution to the Paris Agreement and is most likely not to meet its 2030 emission reduction targets (Climate Action Tracker 2019a). Admittedly, the government did ratify the Paris Agreement in October 2016, but did not vow to higher reductions in greenhouse gas emissions than the previous government, i.e. cutting emissions by 30 percent below 2005 levels until 2030. In addition, the Liberals have actively been supporting the expansion of controversial fossil fuel infrastructure. Most notably, the government has approved the expansion and purchase of the Trans Mountain pipeline, allowing the transport of crude oil from Alberta's oil sands to the coast of British Columbia. Among the government's reasons were to be able to export oil to other markets than the US, to secure jobs and boost economic performance (Tasker 2019). This decision was highly controversial because of potentially compromising the environment and marine life, as well as concerns by Indigenous communities and the impacts the project would have on Canada's greenhouse gas emissions. Indeed, the project stands in stark contrast to the efforts to regulate and reduce greenhouse gas emissions by implementing carbon pricing across Canada.

The Pan-Canadian Framework on Clean Growth and Climate Change (Government of Canada 2016), adopted in 2016, comprises strategies and measures for emission reduction including a carbon pricing plan. The plan sets out a federal benchmark with which all emission pricing systems in the jurisdictions should comply to in order to streamline the approach over the whole economy. Canada's jurisdictions may decide over the system of carbon pricing. They may choose either an explicit price-based system such as a carbon tax or a cap-and-trade-system. If jurisdictions fail to meet the benchmark, the federal government introduces an explicit price-based carbon pricing system to the jurisdiction (Environment and Climate Change Canada 2019b).

... And provincial headwinds

Federal climate action has been facing strong headwinds in the provinces. As of December 2019, the provinces of Alberta, Ontario, New Brunswick, Manitoba and Saskatchewan do not have a carbon pricing system in place that meets the federal benchmark requirements of the federal carbon pricing plan. Alberta cancelled the carbon tax and Ontario cancelled its cap-and-trade system after a change in government (Walker 2020). As a consequence, the federal government introduced a carbon pricing system to four of the provinces, and Alberta is to follow in early 2020. New Brunswick's new carbon pricing plan will replace the imposed federal pricing system in early 2020. As a reaction to the federal level imposing a carbon pricing plan to the provinces, they challenged the constitutionality of the federal carbon pricing system before court. The highest courts in the provinces of Saskatchewan and Ontario ruled in favor of the federal government, but the provincial governments announced they would appeal the decision to the Supreme Court of Canada.

In addition, recent changes in government have made it difficult for the federal government to implement climate change policies. One of the striking examples is oil-rich Alberta, which switched from NDP leadership, the left-leaning social-democratic Canadian party, back to conservative leadership which was aided by the merger of two conservative parties. Alberta had been governed by the center-right Progressive Conservative Association of Alberta for 44 years until 2015. The 2015 provincial elections saw a landslide victory for the social-democratic New Democratic Party. Despite Alberta's long-lasting fossil fuel identity, the NDP provincial government introduced measures to mitigate climate change through a plan endorsed both by environmentalists and the oil sands industry, including a carbon tax, the phasing out of coal-fired electricity and an emphasis on wind power (CBC 2015). Alberta committed to ending emissions from coal-fired electricity and replacing it with 30 percent renewable energy by 2030 (Government of Canada 2016, p. 11). This radical shift of one of the most important fossil fuel and oil sand provinces in Canada has also been interpreted as a reaction to the increasing lack of "social license" to continue oil sand exploitation, which is exemplified by the delay and the protest against each new planned pipeline. In a keynote speech at an event on Canada's climate change policy organized by York University in Toronto in 2015, environmentalist Tzeporah Berman interpreted Alberta's radical shift as a deliberate strategy to create a "green oil" discourse around Alberta's oil sands. If the world continued with oil exploitation at all, then it should be Canadian "green oil" coming along with a carbon tax and an emphasis on renewables (Berman 11/27/2015). Yet, the new Conservative government led by Premier Jason Kenny reversed many of the NDP's environmental policies. It passed the Carbon Tax Repeal Act, which canceled the province's carbon levy, amongst rolling back a number of other climate policies. Nonetheless, Alberta scores third place with regard to installed wind power capacity in Canada and is open to new developments, with 1,483 MW installed capacity as of December 2018 (CanWEA 2019a).

On the other hand, other provinces have been strong proponents of mitigating climate change and ramping up renewable energy. Canada's second largest wind market, after Ontario, is Quebec. 3,882 MW capacity provide for about 5 percent of overall electricity demand (CanWEA 2019a). The province of Ontario has also been regarded as a forerunner in energy system change, phasing out coal and promoting wind energy. Yet, after the July 2018 provincial elections, Ontario has become an example of fierce resistance to federal climate policy. This particularly interesting case will be discussed in more detail later. For the future, it remains to be seen whether "Canada's constitutional separation of (wind) power" (Valentine 2010) will prove obstructive or helpful for reducing greenhouse gas emissions and ramping up renewable energy. The energy-related constitutional division between provinces and federal government has in the past been regarded as a barrier to renewable energy and efficiency programs (Liming et al. 2008) as those provinces that heavily rely on fossil fuel sources may forcefully object to any plans to change the basis of their economy. It also remains to be seen whether the new minority government under Liberal leadership in power since the October 2019 federal elections will put the brakes on or accelerate the implementation of stronger policies on climate change and renewables and balance the provincial resistance.

Ontario: Canada's forerunner in wind energy

The Province of Ontario is an example of a Canadian province that has turned from a provincial leader in climate action to a fierce opponent of federal climate action. The turn was marked by the June 2018 provincial election where the incumbent Liberal government saw a landslide loss from a previous 55 to 7 seats, losing official party status. The Progressive Conservatives won just over 40 percent of the vote (76 seats) and formed a majority government. With incoming Premier Doug Ford, a period of serious setbacks for Ontario's environmental and climate record began. Soon after taking office, he reduced funding for environmental agencies, eliminated the independent office of the environmental commissioner of Ontario and cancelled Ontario's cap-and-trade system for greenhouse gas emissions. The Made-in-Ontario Environment Plan, released in November 2018, aims at reaching climate goals without carbon pricing, but has been heavily criticized for being insufficient (CBC News 2019d; Walker 2020). As the Pan-Canadian Framework specifies, the federal government imposes a carbon levy on any province that fails to meet the minimum federal requirements of a carbon pricing system. The federal government thus introduced such a system to Ontario, which the Progressive Conservatives appealed before court. The provincial government also launched a public relations battle against the federal government that revealed the populist tendencies of Premier Doug Ford (Lachapelle and Kiss 2019). In August 2019, a law was passed that made it mandatory for gas stations to openly display an anti-federal carbon tax sticker showing that the federal carbon tax would add 4.4 cents per litre to the price of gas in 2019, rising to 11 cents a litre in 2022 (CBC News 2019c). The stickers do not mention the rebates available to residents, and in some cases, stickers were placed upon them stating "climate change will cost us more" (CBC News 2019a). A tree-planting program financed by the cap-and-trade-system was also cancelled to cut provincial costs – the federal government however intervened, spending CAD 15 million over four years to save the program (Jones 2019). Renewable energy is also on the cut list. Premier Doug Ford already ran on an election platform promising to repeal the Green Energy Act (Loriggio 2018). Shortly after taking office, the Minister of Energy, Northern Development and Mines Greg Rickford announced the cancellation of more than 750 renewable energy contracts, arguing this would save the province's taxpayers money, as these projects were "unnecessary and wasteful" (McSheffrey 2018). At the end of 2019, it was however revealed that the cancellation costs were CAD 231 million (Gray 2019). Ontarians were not happy with this policy shift against environmental and renewable energy policy. A March 2019 opinion poll revealed that 45 percent of respondents had a negative opinion about what they had heard the government was doing for the environment (Jones 2019). Ford's performance was also found to affect votes in the 2019 provincial elections. Many of the ridings that Ford had won in the provincial elections switched to the Liberals and a poll revealed that 51 percent said they were less likely to vote for the Conservatives due to Ford (CBC News 2019b).

As these examples show, the 2018 elections indeed mark a turning point for climate and environmental policies in Ontario. The Province has seen a stronger period for environmental leadership under the earlier Liberal government during which Ontario's major endeavors in energy system change described in this book occurred. After introducing Ontario from a geographic viewpoint, the remainder of this section focuses on the time of Liberal leadership during which Ontario became a forerunner in wind energy.

Ontario is Canada's most populous province and spreads over a landmass of 918,000 km². The province is highly geographically diverse, with polar bears at home in its frigid northern region and vineyards taking root in the south. While Ontario's north is sparsely populated, the largest share of Ontario's 14.7 million inhabitants is concentrated in southern Ontario, where the biggest Canadian city and provincial capital of Toronto (6.3 million inhabitants) and the Canadian capital city of Ottawa are situated. Ontario's GDP is 857,384 million CAD (2018) and the unemployment rate is 5.6 percent (2019) (Ontario Ministry of Finance 2020). Most of Ontario's economic capacity, which represents 38.9 percent of Canada's total GDP (Ontario Ministry of Finance 2020) is concentrated in the greater Toronto area (GTA). Outside the GTA, there are a number of medium-sized urban centers in the southern parts of Ontario but otherwise, southern Ontario is dominantly rural and its landscape is characterized by farmland, farmyards and single detached houses. Population density in southern Ontario is 86 inhabitants per km².

Ontario's two faces of cosmopolitan urban living in the GTA and placid rural life are also reflected in geographically distinct election outcomes. While the center-left Ontario Liberal Party is predominantly supported by the urban populace, the center-right Progressive Conservative Party of Ontario (PC) usually wins more ridings outside the core GTA and in the rural areas. The social democratic Ontario New Democratic Party, the third relevant party in Ontario provincial politics, is represented in both urban and rural areas, but is of less significance in terms of electoral support. The 2014 provincial elections resulting in a Liberal majority has reinforced this urban/rural election pattern between the Liberal party and the PC (Fisher 2014). The Liberals were governing from the 2003 provincial elections that ended eight years of PC government until the 2018 elections. From 2003 to 2013, Liberal leader Dalton McGuinty was Premier of Ontario, succeeding PC Premier Ernie Evers. In 2013, Kathleen Wynne became Ontario's new Liberal Premier and stayed in office until the 2018 elections.

Ontario's energy policy developments described in this book have been termed "the largest policy experiment to date within North America to decarbonize an electricity system" (Stokes 2013, p. 492) and "one of the most significant energy policy overhauls in Canadian history" (Turner 2015). In April 2015, Ontario became the first jurisdiction in North America to completely eliminate coal from its energy production portfolio, representing the "single largest GHG-reduction initiative in North America" (Government of Canada 2016, p. 10). Over the past decade, Ontario has decreased greenhouse gas emissions of its energy sector by more than 80 percent, replacing its coal-fired power plants with renewables (including hydro-electric), refurbished nuclear and natural gas-fired power plants (IESO 2017b). The successful coal-phase out was largely the result of intense grassroots lobbying led by the Ontario Clean Air Alliance and the Ontario Medical Organisation. The campaign started in 1997, emerging from public concern over pollution and human health. The fact that climate change was not a central concern at the time can be attributed to the fact that this topic did not score high on the public agenda at the time the campaign began. The campaign was effective in that all three major parties promised a coal phase-out in the 2003 provincial elections, though with different timelines (Harris et al. 2015).

Along with its success in phasing out coal, Ontario has also become the leading Canadian province in terms of installed wind power capacity, although this development started relatively late in global comparison. The first wind turbines in the form of single turbines and the first commercial wind park, the Huron Wind Project next to the Bruce Nuclear Generating Station, were set up only in 2002 (Mulvihill et al. 2013). This late initial development gained pace quickly. While in 2003 Ontario had only 15 MW of wind turbines generated by 10 turbines, more than 1,000 wind turbines provided a capacity of 1,950 MW in 2012 (Ministry of Energy 2012). As of December 2018, Ontario had 2,577 wind turbines with a total installed capacity of 5,076 MW, supplying approximately eight percent of the province's electricity demand (CanWEA 2019b). The focus of wind turbine development is southern Ontario (IESO 2016) due to the availability of transmission lines and closer vicinity to consumption centers. Wind power is less developed in the northern parts, although some smaller and medium scale wind energy projects exist as well. Those are mainly developed by Indigenous communities in the context of specific governmental support programs including the Green Energy Act discussed later (Karanasios and Parker 2018). The focus of wind turbine development in this study is southern Ontario.

Ontario's forerunner status in wind energy has only been the result of relatively recent policy endeavors. Historically, Ontario's energy mix was made up of a large percentage of hydro power. Fossil fuels, primarily coal, entered the supply picture in the 1950s and nuclear power about 20 years later (Dorman et al. 2001; Rowlands 2007). Ontario's electricity supply consists of 59 percent nuclear, 23 percent hydro-electric, 10 percent natural gas, 6 percent wind, 2 percent solar and 1 percent biomass (Natural Resources Canada 2017). These numbers show that nuclear power represents the major pillar of Ontario's electricity mix. Containing eight reactors, the Bruce Generating Station is the largest nuclear station in the world. Together with the generating station of Darlington, it provides around 50 percent of the province's electricity needs (Ontario Ministry of Energy 2017, p. 45). Ontario's nuclear supply chain is also of economic importance, supporting 180 companies and providing 60,000 jobs (Ontario Ministry of Energy 2017, p. 45).

Nuclear power in Ontario generally enjoys broad public support in contrast to coal-fired electricity (Hanania et al. 2017). According to the NGO Ontario Clean Air Alliance, which played a central role in the coal phase-out campaign, funding for anti-nuclear campaigns is much more difficult to raise than for other environmental issues (Interview, Jack Gibbons and Angela Bischoff). The 2013 Long Term Energy Plan (LTEP) setting out Ontario's vision for its energy sector states that the then planned refurbishment of the Darlington and Bruce Generating Stations received "strong-province wide support" (Ontario Ministry of Energy 2013, p. 29) during the LTEP consultation process. In summarizing the benefits of nuclear refurbishment, the 2013 LTEP states that it is the most cost-effective generation of electricity, that the existing nuclear generating stations would be located in supportive communities and that nuclear generation produces no greenhouse gas emissions (Ontario Ministry of Energy 2013, p. 29). The revised 2017 LTEP endorses the aim to refurbish a total of ten nuclear units at Darlington and Bruce, which is expected to contribute to a total of CAD 90 billion to Ontario's GDP and increase employment by 14,200 jobs annually (Ontario Ministry of Energy 2017, p. 45).

While nuclear power remains an important component of the Ontario energy mix, the first support policies for renewables including wind, solar and biogas were formulated in the early 2000s. This time was a period of high public salience for environmental issues. Air quality scored especially high among the environmental public policy issues of concern. This was not only a result of a rising number of smog days during summer, but also due to intense engagement of the Ontario Medical Association, a membership organization representing Ontario physicians that had also been active in the campaign to phase out coal (Winfield 2011). In response to this growing discourse that depicted electricity generation as a public health challenge and coal-fired generation plants as causing serious air pollution, the Progressive Conservatives (PC) proposed a renewable portfolio standard for renewables in 2003. It aimed at a one percent share of renewables in the province's electricity demand in 2006, rising by one percent each year to reach eight percent in 2013 (Rowlands 2007). As the PC had traditionally handled electricity solely in terms of "cost" and not as now in terms of other issues such as "health", this represented a major policy shift in Ontario energy policy (Rowlands 2007). In the end the program did not materialize, as the 2003 elections replaced the PC government with a Liberal government. The Liberals had run on a platform promising to phase out coal and develop renewables as well (Rowlands 2007).

In 2004, the Liberals acted on their promise and issued the Request for Proposals program (RfP). Several calls for capacity were issued, which all yielded more offers than planned. In total, the RfP program procured 1,510 MW of wind energy capacity in 15 contracts with private developers (IESO 2015, p. 22) and resulted in the first large-scale wind farms in Ontario. Alongside the introduction of this auction scheme tailored for large-scale commercial investors, the government commissioned a report on the role of feed-in tariffs in the development of renewables. The Ontario Sustainable Energy Association (OSEA),[7] founded in 1999, had launched a campaign for the better inclusion of community aspects in renewable energy procurement and authored the study (Gipe et al. 2005). The report drew on examples from Europe, especially Germany, to outline a possible renewable energy policy geared at small-scale generation below 10 MW. It pro-posed a Standard Offer Contract (another word for feed-in tariff) for renewable energy facilities and emphasized its benefits for the further expansion of renew-able energy, as well as its positive effects on communities, air quality and social acceptance:

> Small renewable energy projects dispersed across the province may offer the benefits of distributed generation, pump more of the project develop-ment dollars through rural economies, increase public awareness and accept-ance of renewable energy, and provide a stable domestic market for renewable energy technology that can spur regional manufacturing industries.
>
> (Gipe et al. 2005, p. 9)

The government largely followed the advice and announced, in 2006, the Renewable Standard Offer Program (RESOP, a feed-in tariff). As this program successfully included OSEA's newly introduced discourse on the beneficial community aspects of renewable energy, it can be regarded as another major policy shift in Ontario's renewable energy policy (Rowlands 2007). The RESOP was the first feed-in tariff in Ontario and provided, for projects under 10 MW, a guaranteed price for 20 years, amounting to 11 cents per kWH for wind, water and biomass, as well as 42 cents per kWH for photovoltaic. When launched, it aimed at procuring 1,000 MW over the next ten years (Yatchew and Baziliaus-kas 2011). The RESOP procured 830 MW of renewable energy (including hydro-electric) which comprised 35 contracts for 285 MW of wind energy capa-city (IESO 2015, p. 22). Despite this success in procuring renewable energy in small-scale projects up to 10 MW, the number of contracts and the correspond-ing amount of procured capacity could have been much higher given that con-tracts had been awarded for almost double the capacity (IEA 2013). This gap between the number of RESOP contracts and its actual procurement rates hints at the implementation problems of the program. Permitting issues and grid con-straints indeed became central drawbacks of RESOP, in spite of the launch of a government-funded program that supported the initial set-up of community pro-jects. According to the founder and executive director of this Community Power Fund, land and grid capacity of the best wind resource sites had already been optioned by large-scale corporations because the RESOP was implemented after the RfP program for large-scale projects was announced (Interview, Deborah Doncaster). Small projects therefore did not have a chance to be allotted grid capacity or land and "it all became a game of begging the private sector to give us some space" (Interview, Deborah Doncaster). Furthermore, despite its ori-ginal intention to foster small-scale renewable facilities, RESOP saw a high con-centration amongst project proponents as some broke up larger projects in order to be eligible for a contract (Yatchew and Baziliauskas 2011). These issues prompted renewable energy advocates to campaign for an improved and larger FIT program, the Green Energy Act.

Ontario's 2009 Green Energy Act

The central legislation to boost wind energy was the Green Energy and Green Economy Act (GEA or Green Energy Act) (McRobert et al. 2016; Stokes 2013). While the act has been changed by minor and major amendments over the course of time, we focus here on its original version. The GEA was enacted in February 2009 and aimed at expanding the rate in renewable energy expansion and providing economic opportunities in the midst of the global economic recession. At its heart was a feed-in tariff (FIT) which was developed later that year. The FIT was subdivided into a microFIT program for projects up to 10 kW of capacity with a simpler project approval process and a regular FIT program for larger projects. The FIT provided a 20-year power purchase agreement for wind, solar, and biomass, and a 40-year power purchase agreement for water power projects. The tariffs paid were 13.5 cent per kilowatt hour on-shore wind, 44.3 to 71.3 for solar and 10.3 to 19.5 for bioenergy, depending on the size of the project (Yatchew and Baziliauskas 2011). Given these relatively generous rates in international comparison, which were soon lowered, the initial response to the FIT was very robust. Within the first twelve months after its launch in October 2009, applications for 15,000 MW of renewable supply were received, which equals approximately 43 percent of Ontario's electricity generating capacity (Yatchew and Baziliauskas 2011).

The discourse of *ecological modernization* was a major driver behind Ontario's renewable energy policy. Speaking to the goals of the act to spur Ontario-based manufacturers, the program also included a domestic content requirement, which prescribed each project that was to be supported under the FIT to contain a defined percentage of domestically manufactured content. For wind power projects, this content was 25 to 50 percent and for solar projects, 40 to 60 percent, depending on when the operation started and the project size (Minister of Energy and Infrastructure 2009). This regulation was later rescinded after the World Trade Organization adopted rulings that this requirement was inconsistent with international trade agreements (IESO 2014). Another measure to spur the creation of manufacturing jobs in the wind and solar industry in Ontario was the "historic agreement" (Minister of Energy and Infrastructure 2010) with the Korean company Samsung to develop 2,500 MW of renewable energy projects outside the regular FIT program in exchange for leveraging manufacturing plants to create employment. This arrangement led to 1,100 MW of installed wind power capacity (IESO 2015, p. 22) and materialized in several large-scale wind farms.

While the introduction of the FIT under the GEA was open with regard to how much capacity was to be procured, targets were fixed in the 2010 Long Term Energy Plan (LTEP). It aimed at a 10,700 MW capacity of wind, solar and bioenergy to be procured and 9,000 MW of hydroelectric power until 2018, while reiterating the commitment to nuclear power which should provide for 50 percent of electricity demand (Ministry of Energy and Infrastructure 2010). Given the fact that the previous RfP program was especially successful in procuring the first large-scale wind farms in Ontario, the question arises as to why the Ontario government decided on a shift towards a FIT and the considerable extension of renewable energy capacity. Which storylines and actors were behind the GEA?

Rationale behind the GEA: economic, environmental and community objectives

Under the impression that the existing programs of RfP and RESOP were not able to adequately procure community-owned energy, a large consortium uniting environmental organizations, the Ontario Federation of Agriculture and an Indigenous[8] energy group formed the Green Energy Act Alliance and called for a larger FIT program for renewable energies, deciding that "the Renewable Energy Sources Act of Germany is really what we needed in Ontario" (MarsDD 2010). The fact that the GEA Alliance included politically influential groups and had a strong relationship with the government led to little opposition during the policy design stage (Stokes 2013).[9] A founder of the GEA Alliance called the enactment of the Green Energy and Economy Act (GEA) an "alignment of stars that happened" (Interview, Deborah Doncaster), referring to the politically determined coal-phase out, market deregulation, the first community and Indigenous power projects having started under the previous program, the new minister George Smitherman stepping in who was open to their ideas, and the successful bid to host the seventh World Wind Energy Association Conference in Ontario. This list illustrates the coincidence of events and developments that occurred during the same time period, including important institutional prerequisites such as Ontario's energy market liberalization of 2002, a successful public display of renewable energy in the form of the international wind energy conference and a responsive government.

The enactment of the GEA can be seen as a merger of two major storylines that are well illustrated by how the then Premier of Ontario Dalton McGuinty introduced the act: "We need those jobs, we need clean electricity and we need to assume our responsibility in the face of climate change" (Ferguson and Ferenc 2009). In the same vein, the Green Energy Act has been described as a "bold series of coordinated actions to enhance economic activity, reduce our impact on the climate, and set Ontario on a course to a greener future" (Ontario Ministry of Municipal Affairs and Housing 2009). The storylines that merged and led to the enactment of the GEA were thus about the economic, environmental and community benefits of an enhanced renewable energy procurement.

The environmental and community-focused organizations within the GEA Alliance saw in an extended FIT a chance to substantially increase the rate of community energy, which the previous programs of RfP and RESOP did not achieve. Community-owned energy generation was seen as a means to achieve a more democratic energy system by giving private individuals the chance to contribute to and participate in renewable energy generation. This was also regarded as leading to vibrant rural communities. The GEA Alliance had thus a strong vision on the community benefits of a FIT scheme and was not primarily driven by industrial interest. This is also evident by the fact that the Canadian Wind Energy Association (CanWEA), an industrial interest group, had not become part of the GEA Alliance because their business model was fully compatible with the earlier RfP program that favored large-scale wind projects (Interview, representative CanWEA). In addition to this community-based motivation of the GEA Alliance, it aimed at substantially extending the use of renewable energy in Ontario in order to prove that it worked on a large dimension and was a worthwhile environmentally and climate friendly alternative for the fossil-fuel based Ontario power sector (Interview, key actor GEAA). Next to these community and environmentally focused arguments, the GEA Alliance also supported economic arguments revolving around the creation of a green industry, which is also explained by the fact that organized labor interests were part of the alliance.

The argument of economic opportunities put forward by the GEA Alliance resonated especially well with the then Minister of Energy and Infrastructure who saw the "economic realities and potential for green innovation" at the heart of the GEA (Smitherman 2009):

> Then, 2008 financial crisis. More than anything else, you need to recognize that there are two drivers behind this policy. The first is the elimination of coal. The second is the sheer risk to Ontario with the prospect of a collapsing North American automotive-industry.
>
> (Interview, George Smitherman)

In the face of the 2008 economic crisis, the GEA was thus regarded as an opportunity to create jobs and for Ontario to become a leader in both green economy-based manufacturing and environmental stewardship. The generation of electricity from renewables was further considered to help fill the electricity supply gap of the planned coal phase-out (Interview, George Smitherman). George Smitherman further affirms that he would not have been successful without focusing on creation of employment. In contrast to Germany, "I didn't have 50 or 60 percent of the population anti-nuclear, therefore already opened to the next solution" (Interview, George Smitherman), thus the elimination of coal was a great opportunity for passing the GEA. In a 2008 tour of European countries and California, the minister witnessed how a FIT could lead to large deployment of renewable energy and job creation. He also met with the then German parliamentarian Hermann Scheer, who had been a key figure in bringing about Germany's Renewable Energy Sources Act and served as a major source of reference during the campaign for the GEA (Anonymous interview, key actor of GEA Alliance).

The implementation of the GEA was welcomed with widespread enthusiasm. While the GEA Alliance called it "a new paradigm for Ontario" (Garcia 2009), the then Minister George Smitherman wrote in a newspaper that "the GEA will revolutionise the very way we create, deliver, conserve, even think about energy" (Smitherman 2009). While the GEA and the design of its FIT scheme were thus inspired by and modelled after the German Renewable Energy Act, the implementation of wind turbine siting was carried out in a fundamentally different way than in Germany. As the planning system for wind energy siting is key to understanding controversies around wind power in Ontario, how was wind power planning in Ontario carried out before and after the GEA?

Implementation: securing provincial objectives

Upon the introduction of the GEA, the procedure for wind turbine planning was changed considerably, which had important consequences on the rise of anti-wind protests. Prior to the changes in the context of the GEA, wind turbine planning was carried out under the two key policy documents of Ontario's policy-led planning system, which are the Planning Act as the overarching Ontario legislation on land use planning and the Provincial Policy Statement, which issues further policy direction. The Provincial Policy Statement of 2005 states that planning authorities shall promote and provide opportunities for the development of renewable energy systems, where feasible. Those shall be "permitted in settlements areas, rural areas and prime agricultural areas in accordance with provincial and federal requirements" (Ontario Ministry of Municipal Affairs and Housing 2005, p. 14). Under this legal framework, a wind power developer's proposal was assessed at the municipal level.[10] The municipality's general goals, long term vision and policies for land use planning are determined in its *official plan*. This plan is complemented by *zoning by-laws*, which specify land use allocation in the municipalities' area in more detail. Some of Ontario's upper-tier municipalities[11] and planning boards have approval authority over lower-tier official plans in place of the Minister of Municipal Affairs and Housing. In both lower-tier and upper-tier municipalities, land use planning issues are negotiated and decided upon in local and regional councils with a prescribed involvement of the public in the form of public hearings.

Under the initial process, when a developer proposed a wind project to a municipality, the municipality needed to assess the extent to which the proposed wind development fit into its official plan and by-laws. The municipality had the authority to change both according to the proposal. If a municipality did not wish to have a wind project developed in their jurisdiction, they could deny the building permit. When the developer did not come to an agreement with a municipality, the case was usually brought before the Ontario Municipal Board, an independent public body whose mission is to uphold the overarching legislation. Thus, although there was room for negotiation and municipal influence, a municipality's decision to object to a project might still have been overturned by the Ontario Municipal Board. While there were cases of municipalities that withheld permits (Anonymous interview with consultancy firm), most of the wind projects developed under RfP and RESOP went smoothly through the municipal application process (Interviews with consultancy firm and CanWEA).

While the previous process of wind turbine planning thus prescribed a central role for the municipality, the GEA exempted renewable energy projects from the requirements under the Planning Act and made them subject to a province-coordinated process of streamlined regulatory approvals (Bues 2018; Walker et al. 2014). Under the new process, project developers must obtain their renewable energy approval (REA) from the Approvals Branch of the Ministry of Environment and Climate Change directly. Acquiring the REA requires a number of environmental and other studies to be submitted to the ministry. The granting of the REA also necessitates documentation of the consultation process carried out with the public. This consultation process consists of at least two mandatory public meetings before which all project details and existing studies need to be shared with the public. The developer is required to document and submit all concerns raised by local groups, individuals, municipalities, Indigenous communities and other public bodies and describe how they were addressed (Government of Ontario 2014). After the project developer has submitted all required studies, the ministry opens the proposal for public consultation, after which the REA is granted to the developer. The REA may include certain conditions which address concerns raised during the consultation process.

Another way in which the Ontario government streamlined the process was the introduction of a minimum distance of wind turbines to the nearest non-participating receptor, e.g. a residential house with no turbine on the property. The exact setback varied according to the number of turbines and sound power level of each turbine determined by a noise study, but was set to a minimum of 550 meters in order not to exceed a sound level of 40 dBA at the point of reception. Although this setback was more restrictive than the distances that were commonly established through municipal planning before the GEA (350–450 meters on average), this minimum setback was considered too conservative by some and far too small by others (Hill and Knott 2010).

The ministry's decision to grant a Renewable Energy Approval to the wind energy proponent may be appealed before the Environmental Review Tribunal (ERT). The ERT is an independent and administrative tribunal established by provincial legislation and holds public hearings to issue decisions on whether a particular undertaking is in line with provincial environmental legislation including the Environmental Protection Act. The ERT reviews the decision to grant a Renewable Energy Approval to a wind energy proponent with regard to two issues. The appellants, who are usually local residents organized in a local anti-wind grassroots organization, have the onus of proving either "(a) serious harm to human health; or (b) serious and irreversible harm to plant life, animal life or the natural environment" (Environmental Review Tribunal 2015). The majority of proposed wind energy projects in Ontario are appealed before the ERT, with the only exception being projects that are very remotely situated (Informal conversation ministry staff member). Only two ERT cases on proposed wind energy projects have so far been won by the appellants. In all other cases, the ERT ruled that they failed to provide expert evidence (ELTO 2015). The fact that only two issues (human health and environment) are deemed acceptable by the government to halt a project represents a further streamlining of the wind turbine approval process. Compared to the situation before the GEA, this process restricts opportunities for local input and the fact that only expert evidence counts before the ERT makes the procedure expensive for common citizens.

Before the GEA, an unwanted project was most often brought before the Ontario Municipal Board, but it was the local council that had to bear the costs of engaging experts.

Despite their limited role in the approvals process as compared to before the GEA, some municipalities in Ontario have still actively embraced wind energy projects. As the Ministry of Energy stated in 2012, many municipalities have both participated in and supported local projects, and almost 20 municipalities "are currently building FIT projects, including Belleville, Kingston, Kitchener, Markham, Waterloo and Welland" (Ministry of Energy 2012, p. 16), but at the same time, the ministry concedes that "there is room for municipalities to play a greater role" (Ministry of Energy 2012, p. 16). While the largest share of wind power projects under the GEA have not been built with municipal involvement, private developers have however often voluntarily engaged with them. One widely adopted measure is the set-up of a "community vibrancy fund" which the developer voluntarily equips with financial resources that the local municipality may use for various purposes of communal concern. Another possibility for community engagement is cooperatively organized wind projects, which however play a minor role in Ontario. While a myriad of small-scale solar and biogas projects was realized under the microFIT,[12] only two wind energy cooperatives exist in Ontario. One is a single showcase turbine in Toronto initiated by the environmental organization TREC and the other is a cooperative in Oxford County that became operational at the end of 2016 (Bowie 2013). Additionally, there are a number of Indigenous community wind farms in Ontario and across Canada (Henderson and Sanders 2017) but as they function within a different legislative and political framework, they are not further considered here.

As community involvement had originally been a primary motivation for the GEA Alliance to call for the act, the question arises as to which of the underlying storylines led to the decision to foster wind energy siting by severely restricting local input. Several interlinked lines of argumentation can be distinguished here. First, the decision was based on prior consultations with the Association of Municipalities in Ontario (AMO) and local municipal representatives (Interviews with key actors of GEA Alliance and with George Smitherman) and the GEA Alliance made efforts to get municipalities on board as another interest group (Anonymous interview with key actor of GEA Alliance). Members of the GEA Alliance convened a meeting at Toronto City Hall with several councilors and mayors from across Ontario. At the meeting, concerns were raised that municipalities would be the first point of conflict in case local residents objected to wind turbine development in their communities, as one of the founding members of the GEA Alliance recalls:

And so they kind of didn't want that responsibility, and so by taking it out of the municipalities' hands and putting it into the province's hands, the municipality can say, "Not my, you've got to go talk to the province". They won't have to mediate that very localized issue. So there was a lot of support within municipalities for that. And but at the same time, that support wasn't particularly vocalized or made public because they don't want to appear to their constituents to be passing the buck so to speak. So there was we felt there was enough support from who we had consulted with that it was ok, plus because it, that was the way that it had been regulated in other jurisdictions, we thought ok, it's fine. A lot of localized policy stuff has been uploaded to the province.

(Interview, Deborah Doncaster)

Lifting the responsibility from the municipalities was thus seen as a way to facilitate project development and help municipalities deal with local opposition. This decision was legitimized by the consulted municipalities, who favored not having the responsibility. At the same time, municipal support for the decision was not made public in order to avoid being blamed by their constituents for voluntarily passing this power to the province. A related second reason to change the siting procedure was that municipalities were not deemed to possess enough capacities and resources to carry out large wind project siting procedures, given that there are a large number of small municipalities in Ontario. The then Minister of Energy and Infrastructure explains:

You can imagine the complexity of a 100 Megawatt wind farm application on a municipality that has one and a half full-time employees and one guy that uses a grader to plough the road. Some of these places are really small. When you go in there, they are one-office towns. So, in some ways their capacities were outstretched, but in the end of the day we uplifted that responsibility, we put a circle around it and said: "It's an Ontario-wide provincial objective". Now, in retrospect, lots of people think that was heavy-handed and alterations have been made, but I continue to defend the earlier instinct.

(Interview, George Smitherman)

As this quote highlights, streamlining the approvals process was regarded in the light of achieving a greater provincial objective and avoid overstretching the resources of municipalities with complex approvals processes. As described earlier, this provincial objective primarily referred to the attraction of foreign investment to create employment and therefore boost Ontario's economy. It was understood that a streamlined decision-making process would make it easier for investors to navigate the application process rather than dealing with numerous small municipalities that each have their own procedure, especially if projects extended over several municipalities (see also Ontario Ministry of Municipal Affairs and Housing 2009). The motivation to attract investment to Ontario is also reflected in the following quote by the then Premier of Ontario, Dalton McGuinty:

To leave decision making, about setbacks for example, to the local authorities would be to effectively keep the [renewable energy] sector out of

Ontario ... it would have driven a stake through the heart of the FIT program before it even got out the door. The local politics are just so difficult to manage. So the [Mayor] says, "it would be good for me and for the municipal finances, but I can understand the neighbours don't need that hassle; we'll just tell them to go elsewhere". And then word gets out that Ontario is not really serious. Because the proponent says, "We've been sent to 7 different municipalities so far and we can't get in because local opposition has mounted and its just not possible". So to be realistic about these things, what you've got to do is have a single, province-wide standard.

(Dalton McGuinty, cited in Stokes (2014, p. 11))

This quote also highlights the third reason for shifting the decision-making power to the province. Dalton McGuinty expressed the concern that local mayors may have been able to halt renewable energy development in Ontario even before the sector could establish itself because local mayors would probably favor their local constituents' viewpoints over the provincial objective. In contrast to the first reason put forward by the former member of the GEA Alliance cited earlier, which depicted the mayors in favor of wind energy projects, this viewpoint regards them as possible opponents. Some decision-makers therefore did not necessarily believe that local communities would have advantages from the planned wind power development which could make them embrace them.

Despite this diverse set of justifications for superseding the municipal level (consultation meetings, limited resources of municipalities, Ontario province-wide objective to attract investment and create employment opportunities) only one explanation reached public debate. In a February 2009 introduction of the GEA, Dalton McGuinty told reporters that only safety issues and environmental standards would be accepted as reasons for objecting to a planned facility, but Ontarians may not say that they just did not like it:

We're going to say to Ontarians that it's okay to object on the basis of safety issues and environmental standards; if you have real concerns there, put those forward and we must find a way to address those, (... ...) But don't say, "I don't want it around here".... NIMBYism will no longer prevail.

(Dalton McGuinty, cited in Ferguson and Ferenc 2009)

This framing of Ontario's way of implementing its renewable energy policy in terms of overcoming selfish Not-In-My-Backyard behavior was the only reason that reached the public. It also conveyed the impression to the public that there was not much to gain from renewable energy projects and that they would be forced upon the local level. By saying "will no longer prevail" in his statement, Dalton McGuinty referred to anti-wind resistance to an off-shore wind energy project at the Scarborough Bluffs, a city district of Toronto.

Summary: Ontario's discursive energy space

Ontario's discursive energy space can be characterized as follows. Ontario enjoys complete sovereignty over its energy sector and can therefore design and change its own energy policies mostly independent from the federal level. This contributes to a weak link to the federal system which is part of the formal institutional context. Ontario's space of participation includes a streamlined approvals process for wind turbine planning which was implemented alongside the introduction of the GEA. The new system superseded the municipal level, establishing the main way of communication between the private wind developer and the ministry level in Toronto. The introduction of the GEA followed the discourses on *ecological modernization* and *democratic pragmatism*. Which precise power effects the discursive energy space had and how this impacted the anti-wind movement's ability to scale-up their concerns will be discussed in a later chapter.

Notes

1 West and East Germany were reunified in 1989. The described development before 1989 refers to West Germany.
2 English: "Energy-turnaround: growth and prosperity without mineral oil and uranium".
3 Municipalities were still allowed to implement smaller setbacks if they included this in their communal plan following the applicable planning rules.
4 All forthcoming quotes regarding Brandenburg are translated from German into English.
5 A *Landkreis* corresponds to a county made up of several municipalities. A non-associated town is not part of a *Landkreis*. In Brandenburg, those are the towns of Brandenburg an der Havel, Cottbus, Frankfurt (Oder) and the state capital of Potsdam.
6 This directive was framed by national legislation including the then version of the Regional Planning Act (ROG: "*Raumordnungsgesetz*") and also anticipated the 1997 change in the National Building Code, which prioritized wind turbines over other land uses unless regional or communal planning is in place. The announcement, however, did not make any reference to those national requirements.
7 OSEA is a non-governmental organization founded to "represent the goals and interests of the sustainable energy and community power sectors" (Garcia 2009). It has strong ties with the renewable energy co-operative TREC which established, in 1999, the first community-owned wind turbine at Toronto's waterfront intended as a show case for community wind projects.
8 In this book, the general term "Indigenous" is used to refer to Indigenous communities of Ontario, including, but not limited to, First Nations and Métis communities (Joseph 2016). The author recognizes that each Indigenous group and person may have their unique preferences of how to be called.
9 The founding organizations of the Green Energy Act Alliance were the Ontario Sustainable Energy Association (OSEA), the Ontario Federation of Agriculture (OFA), the Community Power Fund, Environmental Defence Canada, WWF, the Ivey Foundation, the First Nations Energy Alliance, the Pembina Institute and the David Suzuki Foundation (Garcia 2009).
10 Next to the municipal level, a proponent also required land use planning and environmental approvals from the Ontario government.

11 Those are counties and regional/district municipalities which generally deal with those land use issues that affect more than one municipality. Some of them have their own official plan. There are 30 upper-tier municipalities in Ontario.

12 The microFIT program awarded more than 26,000 contracts which represented over 230 MW (IESO 2017a).

References

AEE (2019a): Anteil Erneuerbarer Energien an der Bruttostromerzeugung – Strom – BB – Daten und Fakten zur Entwicklung Erneuerbarer Energien in einzelnen Bundesländern – Föderal Erneuerbar. Agentur für Erneuerbare Energien. Available online at www.foederal-erneuerbar.de/landesinfo/bundesland/BB/kategorie/strom/auswahl/510-anteil_erneuerbarer_/#goto_510, checked on 12/10/2019.

AEE (2019b): Wichtig für den Kampf gegen den Klimawandel: Bürger*innen wollen mehr Erneuerbare Energien. Agentur für Erneuerbare Energien. Available online at www.unendlich-viel-energie.de/themen/akzeptanz-erneuerbarer/akzeptanz-umfrage/akzeptanzumfrage-2019, checked on 12/20/2019.

AfD Brandenburg (2014): Bodenständig und frei leben. Das Programm der Alternative für Deutschland für die Landtagswahl am 14. September 2014. Available online at www.afd-brandenburg.de/wp-content/uploads/2014/04/Landtagswahlprogramm-komplett.pdf, checked on 8/1/2017.

AfD Brandenburg (2019): Landtagswahlprogramm 2019. Available online at https://afd-brandenburg.de/programm-3/landtagswahlprogramm-2019/, checked on 12/21/2019.

Agentur für Erneuerbare Energien (2012): Brandenburg – Gesamtsieger Leitstern 2012. Berlin. Available online at www.unendlich-viel-energie.de/fileadmin/content/Panorama/Veranstaltungen/LEITSTERN_2012/Factsheets/AEE_LEITSTERN2012_Brandenburg.pdf, updated on 12/12/2012, checked on 3/12/2013.

Agora Energiewende (2018): Stromnetze für 65 Prozent Erneuerbare bis 2030. Zwölf Maßnahmen für den synchronen Ausbau von Netzen und Erneuerbaren Energien. Available online at www.agora-energiewende.de/fileadmin2/Projekte/2018/Stromnetze_fuer_Erneuerbare_Energien/Agora-Energiewende_Synchronisierung_Netze-EE_Netzausbau_WEB.pdf.

Agora Energiewende (2019): Die Energiewende im Stromsektor: Stand der Dinge 2018. Rückblick auf die wesentlichen Enwicklungen sowie Ausblick auf 2019.

Bafoil, François (Ed.) (2016): L'Énergie Éolienne en Europe. Conflits, Démocratie, Acceptabilité Sociale. Paris: Sciences Po les presses (Domaine Développement durable).

Berman, Tzeporah (2015): Keynote Speech at: On the Road to Paris: The Trudeau Government and the Environment, Energy and Climate Change. Organized by York University, Environmental Studies. Toronto, 11/27/2015.

BMWi (2019): Erneuerbare Energien. Bundesministerium für Wirtschaft und Energie (BMWi). Available online at www.bmwi.de/Redaktion/DE/Dossier/erneuerbare-energien.html, checked on 12/10/2019.

Bowie, Tara (2013): ProWind Proposing a Co-Operative Approach to Gunn's Hill Wind Farm. In *Woodstock Sentinel-Review*, 5/8/2013. Available online at www.norwichgazette.com/2013/05/08/prowind-proposing-a-co-operative-approach-to-gunns-hill-wind-farm, checked on 11/9/2015.

Bues, Andrea (2018): Planning, Protest, and Contentious Politics. In *disP – The Planning Review* 54 (4), pp. 34–45. DOI: 10.1080/02513625.2018.1562796.

Busch, Henner; McCormick, Kes (2014): Local Power: Exploring the Motivations of Mayors and Key Success Factors for Local Municipalities to Go 100% Renewable Energy. In *Energy, Sustainability and Society* 4 (1), p. 5. DOI: 10.1186/2192-0567-4-5.

Busch, Per-Olof; Jörgens, Helge (2012): Europeanization through Diffusion? Renewable Energy Policies and Alternative Sources for European Convergence. In Francesc Morata, Israel Solorio (Eds.): European energy policy. An environmental approach. Cheltenham: Edward Elgar, pp. 66–84.

BWE (2016): Windenergie in Brandenburg. Bundesverband WindEnergie e.V. Available online at www.wind-energie.de/en/node/857, checked on 11/21/2016.

Byzio, Andreas; Mautz, Rüdiger; Rosenbaum, Wolf (2005): Energiewende auf schwerer See? Konflikte um die Offshore-Windkraftnutzung. München: Oekom Verlag.

Canada Energy Regulator (2019): Provincial and Territorial Energy Profiles – Canada. Available online at www.cer-rec.gc.ca/nrg/ntgrtd/mrkt/nrgsstmprfls/cda-eng.html, checked on 10/11/2019.

CanWEA (2019a): Installed Capacity – Canadian Wind Energy Association. Available online at https://canwea.ca/wind-energy/installed-capacity/, checked on 9/16/2019.

CanWEA (2019b): Ontario – Canadian Wind Energy Association. Available online at https://canwea.ca/wind-energy/ontario-market-profile/, checked on 9/16/2019.

CBC (2015): Alberta's Climate Change Strategy Targets Carbon, Coal, Emissions. Canadian Broadcasting Corporation 2015, 11/23/2015. Available online at www.cbc.ca/news/canada/edmonton/alberta-climate-change-newser-1.3330153, checked on 4/21/2016.

CBC (2016): 'Canada's Efforts Will Not Cease': Trudeau Signs Paris Climate Treaty at UN, Harness Renewable Energy. Canadian Broadcasting Corporation 2016, 4/24/2016. Available online at www.cbc.ca/news/politics/paris-agreement-trudeau-sign-1.3547822, checked on 6/2/2016.

CBC News (2019a): Climate change sticker parodies Doug Ford's carbon tax sticker. Available online at www.cbc.ca/news/canada/toronto/green-climate-change-sticker-doug-ford-carbon-tax-ontario-1.5118798, checked on 1/21/2020.

CBC News (2019b): Doug Ford's government hurt Andrew Scheer in Ontario, Vote Compass data suggests. Available online at www.cbc.ca/news/canada/toronto/doug-ford-vote-compass-results-1.5329652, checked on 12/9/2019.

CBC News (2019c): New law forcing Ontario gas stations to display anti-carbon tax stickers kicks in | CBC News. Available online at www.cbc.ca/news/canada/toronto/carbon-tax-stickers-law-ontario-1.5265380, checked on 12/9/2019.

CBC News (2019d): Premier Doug Ford says Ontario on track for 2030 emissions targets despite auditor's doubts. Available online at www.cbc.ca/news/canada/toronto/ford-auditor-report-1.5387686, checked on 12/10/2019.

Climate Action Tracker (2019a): Canada | Climate Action Tracker. Available online at https://climateactiontracker.org/countries/canada/, checked on 10/18/2019.

Climate Action Tracker (2019b): Germany | Climate Action Tracker. Available online at https://climateactiontracker.org/countries/germany/, checked on 12/19/2019.

Cuddy, Andrew (2010): Troubling Evidence. The Harper Government's Approach to Climate Science Research in Canada. Climate Action Network Canada. Ottawa.

Deutsche Windguard (2019): Status des Windenergieausbaus an Land in Deutschland. Jahr 2018. Available online at www.windguard.de/jahr-2018.html, checked on 12/19/2019.

Dorman, Arslan; Morrison, Robert W.; Doern, G. Bruce (2001): Canadian Nuclear Energy Policy. Changing Ideas, Institutions, and Interests. Toronto: University of Toronto Press.

ELTO (2015): Hearings – Environment and Land Tribunals Ontario. Environment and Land Tribunals Ontario. Available online at http://elto.gov.on.ca/ert/hearings/, checked on 4/28/2016.

Environment and Climate Change Canada (2019a): Greenhouse gas emissions. Available online at www.canada.ca/en/environment-climate-change/services/environmental-indicators/greenhouse-gas-emissions.html, updated on 10/12/2019, checked on 10/12/2019.

Environment and Climate Change Canada (2019b): Pan-Canadian Approach to Pricing Carbon Pollution. Available online at www.canada.ca/en/environment-climate-change/news/2016/10/canadian-approach-pricing-carbon-pollution.html, updated on 10/23/2019, checked on 10/23/2019.

Environmental Review Tribunal (2015): Decision. Case No.: 15–028 East Oxford Community Alliance Inc. v. Ontario (Environment and Climate Change). Environmental Review Tribunal. Available online at www.ert.gov.on.ca/english/hearings/index.htm, checked on 10/22/2015.

European Commission (2019a): Brandenburg – Internal Market, Industry, Entrepreneurship and SMEs. Available online at https://ec.europa.eu/growth/tools-databases/regional-innovation-monitor/base-profile/brandenburg, checked on 12/21/2019.

European Commission (2019b): Factsheet: Renewable Energy Directive – European Commission. Available online at https://ec.europa.eu/energy/en/content/factsheet-renewable-energy-directive, updated on 12/19/2019, checked on 12/19/2019.

FA Wind (2015): Steuerung der Windenergie im Außenbereich durch Flächennutzungsplan im Sinne des §35 Abs. 3 Satz 3 BauGB. Hintergrundpapier. Fachagentur Wind an Land. Berlin.

FA Wind (2016): Umfrage zur Akzeptanz der Windenergie an Land. Frühjahr 2016. Ergebnisse einer repräsentativen Umfrage zur Akzeptanz der Nutzung und des Ausbaus der Windenergie an Land in Deutschland. Available online at www.fachagentur-windenergie.de/fileadmin/files/Veroeffentlichungen/FA_Wind_Umfrageergebnisse_Fruehjahr_2016.pdf, checked on 6/27/2016.

FAZ (2016): Kommentar: Milliarden in den Wind. In *Frankfurter Allgemeine Zeitung* 2016, 6/3/2016. Available online at www.faz.net/aktuell/wirtschaft/wirtschaftspolitik/die-energiewende-wird-zum-fass-ohne-boden-14264601.html, checked on 6/3/2016.

Federal Ministry for Economic Affairs and Energy (2015): 2014 Renewable Energy Sources Act: Planable. Affordable. Efficient. Available online at www.bmwi.de/EN/Topics/Energy/Renewable-Energy/2014-renewable-energy-sources-act.html, checked on 8/13/2015.

Federal Ministry for Economics Affairs and Energy (2016): For a Future of Green Energy. BMWi. Available online at www.bmwi.de/Redaktion/EN/Dossier/renewable-energy.html, checked on 6/27/2017.

Ferguson, Rob; Ferenc, Leslie (2009): McGuinty Vows to Stop Wind-Farm NIMBYs. In *Toronto Star*, 2/11/2009. Available online at www.thestar.com/news/ontario/2009/02/11/mcguinty_vows_to_stop_windfarm_nimbys.html, checked on 11/2/2015.

Fisher, Robert (2014): Ontario Election 2014: Once again, Kathleen Wynne Settles the Debate. In *CBC*, 6/13/2014. Available online at www.cbc.ca/news/canada/toronto/ontario-votes-2014/ontario-election-2014-once-again-kathleen-wynne-settles-the-debate-1.2674369, checked on 3/20/2017.

Garcia, Roberto (2009): The Green Energy Act – A New Paradigm for Ontario. Presentation at the MMAH 2009 Municipal Symposium, April 30, 2009.

Ge, Mengpin; Friedrich, Johannes; Damassa, Thomas (2014): 6 Graphs Explain the World's Top 10 Emitters. Available online at www.wri.org/blog/2014/11/6-graphs-explain-world%E2%80%99s-top-10-emitters, checked on 4/21/2016.

Gemeinde Schipkau (2016): Gemeinde Schipkau – Energieprojekte der Gemeinde. Available online at www.gemeinde-schipkau.de/texte/seite.php?id=82552, checked on 4/8/2016.

Gipe, Paul; Doncaster, Deborah; MacLeod, David (2005): Powering Ontario Communities: Proposed Policy for Projects up to 10 MW. Prepared by the Ontario Sustainable Energy Association for the Ontario Ministry of Energy. Ontario Sustainable Energy Association. Toronto.

Government of Canada (2016): Pan-Canadian Framework on Clean Growth and Climate Change. Canada's plan to address climate change and grow the economy. Gatineau, Québec: Environment and Climate Change Canada. Available online at www.canada.ca/content/dam/themes/environment/documents/weather1/20170125-en.pdf, checked on 6/19/2017.

Government of Ontario (2014): Environmental Registry: Decision on the Niagara Region Wind Corporation proposal. Available online at www.ebr.gov.on.ca/ERS-WEB-External/displaynoticecontent.do?noticeId=MTIxMTM5&statusId=MTg2MDY0, checked on 6/22/2015.

Gray, Jeff (2019): Ontario Premier Doug Ford defends $231-million cost of killing green-energy deals. The Globe and Mail. Available online at www.theglobeandmail.com/canada/article-ontario-premier-doug-ford-defends-231-million-cost-of-killing-green/, checked on 1/21/2020.

Hanania, Jordan; Jenden, James; Lloyd, Ellen; Donev, Jason (2017): Canadian Support for Nuclear Power. Energy Education. Available online at http://energyeducation.ca/encyclopedia/Canadian_support_for_nuclear_power#cite_note-pfukuCNA-3, updated on 7/28/2017, checked on 12/6/2017.

Harris; Melissa; Beck; Marisa; Gerasimchuk; Ivetta (2015): The End of Coal: Ontario's Coal Phase-Out. International Institute for Sustainable Development. Winnipeg, Manitoba. Available online at www.iisd.org/sites/default/files/publications/end-of-coal-ontario-coal-phase-out.pdf, checked on 5/13/2016.

Hehn, Nina; Miosga, Manfred (2015): Die Zukunft der Windenergie in Bayern nach Einführung der 10 H-Regel. In Bundesinstitut für Bau-, Stadt- und Raumforschung (Ed.): Ausbaukontroverse Windenergie (Informationen zur Raumentwicklung, Heft 6.2015), pp. 631–644.

Henderson, Chris; Sanders, Cara (2017): Powering Reconciliation. A Survey of Indigenous Participation in Canada's Growing Clean Energy Economy. Lumos Clean Energy Advisors. Ottawa.

Hill, Stephen; Knott, James (2010): Too Close for Comfort: Social Controversies Surrounding Wind Farm Noise Setback Policies in Ontario. In *Renewable Energy Law & Policy Rev*iew (153), pp. 153–168.

Hooker, Clifford A.; MacDonald, Robert; van Hulst, Robert; Victor, Peter (1981): Energy and the Quality of Life: Understanding Energy Policy. Toronto: University of Toronto Press.

IEA (2013): IEA – Canada. Ontario Renewable Energy Standard Offer Programme (RESOP). International Energy Agency. Available online at www.iea.org/policiesandmeasures/pams/canada/name-24400-en.php, checked on 7/31/2017.

IESO (2014): Changes Relating to Domestic Content Requirements for microFIT after July 25, 2014. Independent Electricity System Operator. Available online at http://microfit.powerauthority.on.ca/faqs/microfit-domestic-content-after-25-july-2014, checked on 9/28/2016.

IESO (2015): Progress Report on Contracted Electricity Supply. Second Quarter 2015. Independent Electricity System Operator. Available online at www.ieso.ca/-/media/

Files/IESO/Document%20Library/contracted-electricity-supply/Progress-Report-Contracted-Supply-Q22015, checked on 11/20/2016.

IESO (2016): Ontario's Electricity System. Independent Electricity System Operator. Available online at www.ieso.ca/ontarioenergymap/index.html, updated on 4/13/2016, checked on 4/21/2016.

IESO (2017a): 50 MW Procurement Target Reached – December 1, 2017. Independent Electricity System Operator. Available online at www.ieso.ca/en/get-involved/microfit/news-bi-weekly-reports/50-mw-procurement-target-reached--december-1-2017, checked on 12/5/2017.

IESO (2017b): Ontario's Supply Mix. Ontario's Energy Capacity. Independent Electricity System Operator. Available online at www.ieso.ca/learn/ontario-supply-mix/ontario-energy-capacity, checked on 4/4/2017.

Ipsos Reid (2010): Canadians and Americans Call for More Action on the Environment. Available online at http://angusreidglobal.com/wp-content/uploads/2010/07/2010.07.19_Environment1.pdf, checked on 4/21/2016.

Jones, Allison (2019): Ottawa Pledges to Spend $15 million to Save Ontario's Tree-planting Program. In *The Canadian Press*. Available online at https://globalnews.ca/news/5354025/ottawa-pledges-funding-ontario-tree-planting-program/?utm_expid=.kz0UD5JkQOCo6yMqxGqECg.0&utm_referrer=https%3A%2F%2Fglobalnews.ca%2Fnews%2F5960379%2Ftrudeau-plant-trees-climate-change%2Fbeta%2F, checked on 12/9/2019.

Jones, Allison (2019): Ontarians have Negative Opinion of Ford's Environment Plan. Poll. City News, The Canadian Press. Available online at https://toronto.citynews.ca/2019/03/31/ontarians-have-negative-opinion-of-fords-environment-plan-poll/, updated on 9/26/2019, checked on 9/26/2019.

Joseph, Bob (2016): OPINION | Indigenous or Aboriginal: Which is Correct? CBC News. Available online at www.cbc.ca/news/indigenous/indigenous-aboriginal-which-is-correct-1.3771433, checked on 1/8/2020.

Karanasios, Konstantinos; Parker, Paul (2018): Technical Solution or Wicked Problem? In *Journal of Enterprising Communication* 12 (3), pp. 322–345. DOI: 10.1108/JEC-11-2017-0085.

Kemfert, Claudia (2017): Germany Must Go Back to its Low-carbon Future. In *Nature* 549 (7670), pp. 26–27. DOI: 10.1038/549026a.

Krause, Florentin; Bossel, Hartmut; Müller-Reißmann, Karl-Friedrich (1980): Energie-Wende. Wachstum und Wohlstand ohne Erdöl und Uran; ein Alternativ-Bericht des Öko-Instituts, Freiburg. Frankfurt am Main: Fischer.

Kunze, Sebastian (2015): Von der Theorie in die Praxis: Kommunale Beteiligungen an Windkraftprojekten. Power Point Präsentation. Städte- und Gemeindebund Brandenburg.

Lachapelle, Erick; Kiss, Simon (2019): Opposition to Carbon Pricing and Right-wing Populism. Ontario's 2018 General Election. In *Environmental Politics* 28 (5), pp. 970–976. DOI: 10.1080/09644016.2019.1608659.

Landesregierung Brandenburg (2002): Energiestrategie 2010. Der energiepolitische Handlungsrahmen des Landes Brandenburg bis zum Jahr 2010.

Landtag Brandenburg (2006): Beschluss: Zukunft sichern – Brandenburg als Energieland ausbauen. LT-DS 4/2893-B, Mai 2006.

Landtag Brandenburg (2014): Antwort der Landesregierung auf die Kleine Anfrage 3409 des Abgeordneten Christoph Schulze, Fraktion Bündnis 90/Die Grünen. Drucksache 5/8731. Entwicklung der Windkraft im Land Brandenburg.

Landtag Brandenburg (2019): Beschlossene Gesetze der 6. Wahlperiode – Landtag Brandenburg. Landtag Brandenburg. Available online at www.landtag.brandenburg.de/ de/parlament/plenum/gesetzgebung/beschlossene_gesetze_der_6._wahlperiode/658313, checked on 1/10/2020.

Lauber, Volkmar (2012): Wind Power Policy in Germany and the UK. Different Choices Leading to Divergent Outcomes. In Joseph Szarka (Ed.): Learning from Wind Power. Governance, Societal and Policy Perspectives on Sustainable Energy. Basingstoke: Palgrave Macmillan (Energy, climate and the environment series), pp. 38–60.

Liming, Huang; Haque, Emdad; Barg, Stephan (2008): Public Policy Discourse, Planning and Measures toward Sustainable Energy Strategies in Canada. In *Renewable and Sustainable Energy Reviews* 12 (1), pp. 91–115.

Loriggio, Paola (2018): Ontario Government Moves to Scrap Green Energy Act. CTV News. Available online at www.ctvnews.ca/canada/ontario-government-moves-to-scrap-green-energy-act-1.4102549, checked on 9/26/2019.

LUGV (2016): Klimagasinventur. Eine Bilanz, die über die energiebedingten CO2-Emissionen hinausgeht. Available online at www.lugv.brandenburg.de/cms/detail.php/ bb1.c.296626.de, checked on 9/10/2015.

Marcotullio, P. J.; L. Bruhwiler; S. Davis; J. Engel-Cox; J. Field; C. Gately et al. (2018): Chapter 3: 21 Energy Systems. In Cavallaro, N., G. Shrestha, R. Birdsey, M. A. Mayes, R. G. Najjar, S. C. (Ed.): Second State of the Carbon Cycle Report 23 (SOCCR2): A Sustained Assessment Report. Washington, DC, USA.

MarsDD (2010): GEAFF – Perspectives on the Ontario Green Energy Act – Deborah Doncaster. MaRS Discovery District (Director). Available online at https://vimeo. com/9551136.

McRobert, David; Tennent-Riddell, Julian; Walker, Chad (2016): Ontario's Green Economy and Green Energy Act: Why a Well-Intentioned Law is Mired in Controversy and Opposed by Rural Communities. In *RELP* 2, pp. 1–22.

McSheffrey, Elizabeth (2018): Ontario Cancelling 758 'Unnecessary and Wasteful' Renewable Energy Contracts. *Canada's National Observer*. Available online at www. nationalobserver.com/2018/07/13/news/ontario-cancelling-758-unnecessary-and-wasteful-renewable-energy-contracts, checked on 9/26/2019.

Mecklenburg Vorpommern (2016): Gesetz über die Beteiligung von Bürgerinnen und Bürgern sowie Gemeinden an Windparks in Mecklenburg-Vorpommern (Bürger- und Gemeindenbeteiligungsgesetz – BüGembeteilG M-V) Vom 18. Mai 2016. Dienstleistungsportal Mecklenburg-Vorpommern. Available online at www.landesrecht-mv.de/ jportal/portal/page/bsmvprod.psml?showdoccase=1&st=lr&doc.id=jlr-WindPB%C3% BCGemBGMVrahmen&doc.part=X&doc.origin=bs, checked on 11/6/2017.

Minister of Energy and Infrastructure (2009): Directive to the Ontario Power Authority Sep 24, 2009. Minister of Energy and Infrastructure, Office of the Deputy Premier. Toronto.

Minister of Energy and Infrastructure (2010): Directive to the Ontario Power Authority, May 7, 2010.

Ministry of Energy (2012): Ontario's Feed-in Tariff Program. Two-Year Review Report.

Ministry of Energy and Infrastructure (2010): Long Term Energy Plan. Queen's Printer for Ontario. Toronto.

MIR; MLUV (2009): Hinweise an die Regionalen Planungsgemeinschaften zur Festlegung von Eignungsgebieten „Windenergie". Windkrafterlass Brandenburg. Ministerium für Infrastruktur und Raumordnung des Landes Brandenburg (MIR); Ministerium für Ländliche Entwicklung, Umwelt und Verbraucherschutz (MLUV). Potsdam.

MLUR (2000): Regionalplanung steuert Nutzung der Windkraft. Ministerium für Land-wirtschaft, Umweltschutz und Raumordnung. In *Brandenburger Agrar- und Umwelt-journal* 2 (5), pp. 8–10.

MUGV (2011): Beachtung naturschutzfachlicher Belange bei der Ausweisung von Wind-eignungsgebieten und bei der Genehmigung von Windenergieanlagen. Erlass des Min-isteriums für Umwelt, Gesundheit und Verbraucherschutz vom 01. Januar 2011. Ministerium für Umwelt, Gesundheit und Verbraucherschutz.

Mulvihill, Peter; Winfield, Mark; Etcheverry, Jose (2013): Strategic Environmental Assessment and Advanced Renewable Energy in Ontario. Moving Forward or Blowing in the Wind? In *Journal of Environmental Assessment and Policy Management* 15 (02), 1–19. DOI: 10.1142/S1464333213400061.

MUNR (1994): Windkraftnutzung in Brandenburg. Ministerium für Umwelt, Naturschutz und Raumordnung des Landes Brandenburg. In *Brandenburger Umweltjournal* 4 (2), p. 14.

MUNR (1995): Größter Windpark Brandenburgs am Netz. Ministerium für Umwelt, Naturschutz und Raumordnung des Landes Brandenburg. In *Brandenburger Umwelt-journal* 5 (17), p. 19.

MUNR (1996a): Brandenburger Energiekonzept. Ministerium für Umwelt, Naturschutz und Raumordnung des Landes Brandenburg. In *Brandenburger Umweltjournal* 6 (20), p. 27.

MUNR (1996b): Erlaß des Ministeriums für Umwelt, Naturschutz und Raumordnung zur landesplanerischen und naturschutzrechtlichen Beurteilung von Windkraftanlagen im Land Brandenburg (Windkrafterlaß des MUNR). Ministerium für Umwelt, Naturschutz und Raumordnung Landes Brandenburg (MUNR). Available online at http://gl.berlin-brandenburg.de/imperia/md/content/bb-gl/regionalplanung/windkrafterlass.pdf, checked on 3/14/2013.

MUNR (1996c): Windkrafterlaß für Interessenausgleich von Naturschutz und Umwelts-chutz. Ministerium für Umwelt, Naturschutz und Raumordnung des Landes Branden-burg. In *Brandenburger Umweltjournal* 6 (20), p. 9.

MUNR (1997): Erneuerbare Energie: 5 Prozent bis 2010. Ministerium für Umwelt, Natur-schutz und Raumordnung des Landes Brandenburg. In *Brandenburger Umweltjournal* 8 (26), pp. 26–27.

MW (2008): Energiestrategie 2020 des Landes Brandenburg. Ministerium für Wirtschaft des Landes Brandenburg. Potsdam. Available online at www.energie.brandenburg.de/media_fast/bb1.a.2755.de/Energiestrategie_2020.pdf, checked on 8/28/2012.

MWE (2012): Energiestrategie 2030 des Landes Brandenburg. Potsdam.

MWE (2016): 6. Monitoringbericht zur Energiestrategie des Landes Brandenburg. Berichtsjahr 2014 mit energierelevanten Daten. Ministerium für Wirtschaft und Energie (MWE) des Landes Brandenburg. Potsdam. Available online at www.zab-energie.de/de/system/files/media-downloads/6._monitoringbericht_zur_energiestrategie_-_berichtsjahr_2014_1.pdf, checked on 5/6/2016.

NABU Brandenburg; BUND Brandenburg (2010): Naturschutzfachliche Kriterien für Windeignungsgebiete und für Genehmigungsverfahren von Windkraftanlagen. Avail-able online at http://brandenburg.nabu.de/imperia/md/content/brandenburg2/windkraft-kriterien-stellungnahme_-_april_2010.pdf, updated on 4/21/2010, checked on 10/5/2012.

Natural Resources Canada (2017): Electricity facts. Available online at www.nrcan.gc.ca/energy/facts/electricity/20068, updated on 12/27/2017, checked on 12/28/2017.

Natural Resources Canada (2019a): Crude oil facts. Available online at www.nrcan.gc.ca/crude-oil-facts/20064, checked on 10/9/2019.

Natural Resources Canada (2019b): Renewable energy facts. Available online at www. nrcan.gc.ca/science-data/data-analysis/energy-data-analysis/energy-facts/renewable-energy-facts/20069#L2, updated on 10/12/2019, checked on 10/12/2019.

Ohlhorst, Dörte (2009): Windenergie in Deutschland. Konstellationen, Dynamiken und Regulierungspotenziale im Innovationsprozess. Wiesbaden: VS Verlag für Sozialwissenschaften/GWV Fachverlage GmbH, Wiesbaden (SpringerLink: Bücher).

Ontario Ministry of Energy (2013): Achieving Balance – Ontario's Long-Term Energy Plan. Queen's Printer for Ontario. Toronto.

Ontario Ministry of Energy (2017): Delivering Fairness and Choice. 2017 Long-Term Energy Plan. Queen's Printer for Ontario. Toronto.

Ontario Ministry of Finance (2020): Ontario Fact Sheet. Available online at www.fin.gov. on.ca/en/economy/ecupdates/factsheet.pdf, checked on 1/21/2020.

Ontario Ministry of Municipal Affairs and Housing (2005): Provincial Policy Statement. Queen's Printer for Ontario. Toronto. Available online at www.mah.gov.on.ca/ Page1485.aspx, checked on 8/3/2017.

Ontario Ministry of Municipal Affairs and Housing (2009): Need for Change. Available online at www.mah.gov.on.ca/Page6672.aspx, checked on 8/3/2017.

Overwien, Petra; Groenewald, Ulrike (2015): Viel Wind um den Wind. Aktuelle Herausforderungen für die Regionalplanung in Brandenburg. In Bundesinstitut für Bau-, Stadt- und Raumforschung (Ed.): Ausbaukontroverse Windenergie (Informationen zur Raumentwicklung, Heft 6.2015), pp. 603–618.

OVG Berlin-Brandenburg (2010): Unwirksamkeit des Regionalplans Havelland-Fläming – Sachlicher Teilplan „Windenergienutzung" vom 2. September 2004. Oberverwaltungsgericht Berlin-Brandenburg (OVG B-B), updated on 9/14/2010, checked on 1/28/2014.

Prime Minister of Canada (2015): Canada's National Statement at COP21. Address by the Right Honourable Justin Trudeau, Prime Minister of Canada, Paris, France – 30 November 2015.

Rowlands, Ian (2007): The Development of Renewable Electricity Policy in the Province of Ontario: The Influence of Ideas and Timing. In *Review of Policy Research* Vol 24 (3).

Scharper, Stephen Bede (2015): Climate of Hope Returns. In *Toronto Star* 2015, 11/10/2015. Available online at http://torontostar.newspaperdirect.com/epaper/viewer. aspx, checked on 4/20/2016.

Shah, Maryam (2019): Climate Change Emerges as One of the Top Ballot-box Issues Among Voters. Ipsos Poll. Available online at https://globalnews.ca/news/6006868/ climate-change-federal-election-issue-poll/, checked on 10/12/2019.

Smitherman, George (2009): The Green Revolution. In *Windsor Star*, 3/17/2009. Available online at www.pressreader.com, checked on 11/10/2015.

Solorio, Israel; Öller, Eva; Jörgens, Helge (2014): The German Energy Transition in the Context of the EU Renewable Energy Policy. A Reality Check! In Achim Brunnengräber, Maria Rosaria DiNucci (Eds.): Im Hürdenlauf zur Energiewende. Von Transformationen, Reformen und Innovationen. Wiesbaden: Springer VS (SpringerLink: Bücher), pp. 189–200.

SPD Brandenburg; DIE LINKE Brandenburg (2014): Sicher, Selbstbewusst und Solidarisch: Brandenburgs Aufbruch vollenden. Koalitionsvertrag zwischen SPD Brandenburg und DIE LINKE Brandenburg für die 6. Wahlperiode des Brandenburger Landtages 2014 bis 2019.

SRU (2017): Start Coal Phaseout Now. Statement. German Advisory Council on the Environment. Available online at www.umweltrat.de/SharedDocs/Downloads/EN/04_ Statements/2016_2020/2017_10_statement_coal_phaseout.html, checked on 12/19/2019.

Stokes, Leah C. (2013): The Politics of Renewable Energy Policies: The Case of Feed-In Tariffs in Ontario, Canada. In *Energy Policy* 56, pp. 490–500. DOI: 10.1016/j. enpol.2013.01.009.

Stokes, Leah C. (2014): Electoral Backlash against Climate Policy: A Natural Experiment on Accountability and Local Resistance to Public Policy. Paper presented at annual workshop of the Ontario Sustainable Energy Policy Group (ONSEP).

SZ (2016): Energiewende – Deutschland legt seiner Energiewende Fesseln an. In *Süddeutsche Zeitung*, 2016 (1.6.2016). Available online at www.sueddeutsche.de/wirtschaft/energiewende-deutschland-legt-seiner-energiewende-fesseln-an-1.3014862, checked on 6/3/2016.

Tasker, John Paul (2019): Trudeau Cabinet Approves Trans Mountain Expansion Project | CBC News. Available online at www.cbc.ca/news/politics/tasker-trans-mountain-trudeau-cabinet-decision-1.5180269, checked on 10/12/2019.

The Guardian (2011): Canada Pulls out of Kyoto Protocol. In *The Guardian* 2011, 12/13/2011. Available online at www.theguardian.com/environment/2011/dec/13/canada-pulls-out-kyoto-protocol, checked on 4/20/2016.

Thüringer Ministerium für Umwelt, Energie und Naturschutz (2016): Medieninformation: Siegesmund: Kommunen und Bürger am Ausbau der Windenergie beteiligen. 7 Windanlagenprojektierer erhalten Siegel für „Faire Windenergie". Available online at www.thueringen.de/th8/tmuen/aktuell/presse/89929/index.aspx, updated on 3/21/2016, checked on 11/6/2017.

TNS Emnid (2016): Umfrageergebnisse Emnid EEG. Available online at www.greenpeace.de/sites/www.greenpeace.de/files/publications/20160414_umfrage_emnid_eeg.pdf, checked on 4/29/2016.

Turner, Chris (2015): Tilting at Windmills. Canada's Most Ambitious Green Energy Plan Failed – But Not for the Reason Everybody Thinks. In *The Walrus Magazine*, 2015 (November 2015). Available online at http://thewalrus.ca/tilting-at-windmills/, checked on 10/30/2015.

Valentine, Scott Victor (2010): Canada's Constitutional Separation of (Wind) Power. In *Energy Policy* 38 (4), pp. 1918–1930.

Walker, Chad (2020): Bill 4 and the Removal of Cap and Trade. A Case Study of Carbon Pricing, Climate Change Law, and Public Participation in Ontario, Canada. In *Journal of Environmental Law and Practice* 33 (1), 35–72.

Walker, Chad; Baxter, Jamie; Ouellette, Danielle (2014): Beyond Rhetoric to Understanding Determinants of Wind Turbine Support and Conflict in Two Ontario, Canada Communities. In *Environment and Planning* 46 (3), pp. 730–745. DOI: 10.1068/a130004p.

Wehrmann, Benjamin (2019): Limits to Growth: Resistance against Wind Power in Germany. Available online at www.cleanenergywire.org/factsheets/fighting-windmills-when-growth-hits-resistance, updated on 1/17/2020, checked on 1/17/2020.

Winfield, Mark (2011): Blue-Green Province: The Environment and the Political Economy of Ontario. Ontario: UBC Press. Available online at http://books.google.de/books?id=ac2cvGbePlcC.

Yatchew, Adonis; Baziliauskas, Andy (2011): Ontario Feed-In-Tariff Programs. In *Energy Policy* 39 (7), pp. 3885–3893.

ZAB (2014): 4. Monitoringbericht zur Energiestrategie des Landes Brandenburg. Berichtsjahr 2012 mit energierelevanten Daten. Zukunftsagentur Brandenburg (ZAB). Potsdam.

4   Larger setbacks, saving the forests

The anti-wind movement in Germany, case study Brandenburg

The federal state of Brandenburg is among the forerunners of wind energy deployment in Germany. It won the bi-annual lodestar award on three consecutive occasions and accomplished great achievements in terms of renewable energy political support. In the lodestar evaluation of 2014, however, Brandenburg fell to fifth place. One reason is that other federal states have caught up in renewable energy development, but the Ministry of Economic Affairs and Energy also put this into context with declining levels of social acceptance for renewable energies (MWE 2014). This chapter presents the rise of the anti-wind movement in Brandenburg and how they were received by the opposition party and responded to by the government. A case study of a local anti-wind protest group is presented, including their organization and arguments. The chapter draws on empirical research in Brandenburg carried out in 2016, encompassing 16 interviews with representatives from government, associations, wind companies and politicians at the township and provincial level as well as two focus group discussions with anti-wind groups at the local level. The chapter focuses on the time period from around 2010 to 2016.

While the number of wind turbines has gradually increased in Brandenburg, it was not until the years 2010–2012 that opposition to wind power reached significant levels (Multiple interviews with representatives from civil society, companies and politicians in Brandenburg). A 2009 opinion poll comparing all German federal and city states revealed that a large majority of 93 percent of the Brandenburg population were in favor of further deployment of renewable energy, but only 65 percent were in favor of renewable energy facilities in their neighborhood, which made Brandenburg score last in German comparison (Forsa 2009). While solar panels were still deemed acceptable by 75 percent, wind turbines (44 percent) and biogas plants (39 percent) scored particularly low in acceptance (Forsa 2009). These numbers have since gradually decreased further. A 2016 survey showed that the percentage of the Brandenburg population in favor of further development of renewable energy had dropped to 57 percent, while 43 percent would approve and 38 percent oppose a wind turbine being constructed in their neighborhood (MAZ 2016a).

The reasons for opposition in Brandenburg are mixed and little research has so far explicitly focused on Brandenburg (e.g. Bafoil 2016). A 2015 survey revealed that wind opponents in Brandenburg are more likely to change their mind under certain conditions as compared to opponents in two other federal states (PIK et al. 2016). If no endangered species were threatened, if it was ensured that no risks for human health occurred, if the municipality or they financially benefitted, or if wind turbines were effective against climate change, wind opponents in Brandenburg would be more willing to change their view than the respondents in Baden-Württemberg and Schleswig-Holstein. The survey hints at the importance of endangered species, health risks and financial participation as topics of concern in Brandenburg. The variety of anti-wind arguments is also visible within the formal wind power planning system and the general public debate in Brandenburg.

Protest within the formal planning system and public debate in Brandenburg

The first generation of Brandenburg's regional plans that entered into force in the mid-2000s did not legally require public consultation and the number and scope of comments made by municipalities and other consulted legal bodies remained limited (Overwien and Groenewald 2015). For the development of the subsequent generation of regional plans, public consultation as an inherent part of the process required that each draft version prepared by the regional planning authorities (RPGs) be subject to a public commenting period. Comments submitted by municipalities, affected public bodies, organizations, companies and individuals need to be considered by the planning offices of the RPGs and be included in a revised version of the regional plan or be dismissed on legitimate grounds. In the context of this public consultation process, the RPGs usually receive thousands of comments by individual citizens and not all RPGs make them publicly available. One exception is the RPG of Havelland-Fläming, whose public consultation results is presented here to illustrate the scope and content of the submitted comments by individual citizens.

In two distinct public consultation processes for the first and second draft of the regional plan in 2013 and 2014, the RPG received 5,801 comments by individual citizens including 94,924 points of concern and suggestions (E-mail correspondence with staff member RPG Havelland-Fläming). The four main submitted issues were human well-being, biodiversity, energy policy and economic considerations (see Table 4.1), but other matters were also raised. An analysis of these numbers however must consider the fact that many of the submitted comments stemmed from serial letters that made use of prepared text modules provided on citizens' initiatives websites (Interview, staff member RPG Havelland-Fläming). Still, they provide an impression of the topics raised during a typical public consultation process in Brandenburg. As Table 4.1 highlights, danger of forest fire constitutes the single largest concern under the issue area of human well-being. This highlights the regional specificity of raised concerns. Due to the fact that in the RPG of Havelland-Fläming, the proposed regional plans devised a number of wind designation areas in

Table 4.1 Comments by citizens in two participation processes, compiled and clustered by the RPG of Havelland Fläming

Issue area and selected sub-aspects	Number of comments
Human well-being (health, setbacks, quality of life, risk of fire)	27,411
Danger through sound/engine and construction noise	2,854
Infrasound and low-frequency sound	2,764
Danger of forest fire	7,214
Flora, fauna, biological diversity in general	20,763
Energy policy	12,786
Particular species	9,100
Cultural and material goods	5,726
Loss in property value	2,937
Economic efficiency of wind turbines	4,613
Economy (negative impacts on tourism)	4,357
Landscape	3,242
Miscellaneous	2,362
Climate/air	1,846
Water	1,446
Soil	960
Wind turbine relocation and repowering	318

Source: E-Mail Communication with Staff Member.

existing forests, local protest has mainly revolved around the fear of turbine-induced forest fire.

Submitted comments in the context of public consultation processes in Brandenburg usually also contain positive statements and requests to enlarge proposed wind suitability areas or to increase their number. In the second public consultation process of Havelland-Fläming, companies and land owners asked for 27,000 hectares of additional area for wind suitability areas and were supported by ten law firms (RPG Havelland-Fläming 6/17/2014).

As illustrated by this public consultation process, comments submitted in the context of the official consultation process span from very general concerns such as energy policy to very specific concerns such as a possible threat to one particular species. Yet, active engagement against wind turbines in the form of participating in the official consultation process remains a rather local issue in Brandenburg. According to the Brandenburg state planning agency, the majority of Brandenburg's population does not participate in the regional planning process as the number of comments received, which also include proponents of wind energy, does not correspond to more than one percent of the population of one RPG (Overwien and Groenewald 2015). This is, however, not surprising as urban residents not directly affected by the proposed wind turbines are probably not inclined to submit negative comments. Nonetheless, the discussion on wind energy has reached strong intensity, which has also become visible during the public meetings of the regional assemblies, especially when the discussion of draft versions of the regional plan is on the agenda. It is common that a large number of local anti-wind opponents join the meetings showing protest signs and heckling the meeting.[1]

Scaling-up the debate: The Volksinitiative "Rettet Brandenburg"

As of July 2014, more than 80 citizens' initiatives[2] against wind energy were counted in Brandenburg (Jäger 2015). Many of them are organized under the Brandenburg-wide campaign *Volksinitative Rettet Brandenburg* (popular initiative Save Brandenburg), which has contributed significantly in scaling-up localized anti-wind disputes to the state level. The legal basis of *Rettet Brandenburg's* popular initiative is as follows. All residents in Brandenburg are entitled to initiate a popular initiative (*Volksinitiative*) of which the aim is to have the state parliament address a specific topic. If at least 20,000 residents sign the petition, state parliament must reach a decision over its demands within four months. If state parliament declines them, the initiators may proceed to the second step of a public referendum (*Volksbegehren*) which requires 80,000 residents to submit their signatures to the offices of the local authorities (Landtag Brandenburg 2009). A previously successful referendum that serves as a frequent reference for wind opponents submitted more than 100,000 signatures against large-scale livestock breeding farms in Brandenburg in 2016 (Mehr Demokratie 2017).

The first anti-wind popular initiative was initiated by a group of 18 citizens' initiatives in 2008 called *Gegen die Massenbebauung Brandenburgs mit Windenergieanlagen!* (Against the mass development of wind turbines in Brandenburg!). In April 2009, they submitted 27,171 signatures to state parliament (Landtag Brandenburg 2015d). Their central demands included a general setback of 1,500 meters between wind turbines and residential areas, a setback of 10 km between two wind turbines and that wind turbines be placed in former military areas and former open cast mines instead of nature protection areas (Tagesspiegel 2009). State parliament however dismissed their demands. The initiators at the time did not proceed to the second step of a referendum.

In 2015, *Rettet Brandenburg* (Save Brandenburg) started a new popular initiative called *Volksinitiative für größere Mindestabstände von Windrädern sowie keine Windräder im Wald* (People's initiative for greater minimum setbacks of wind turbines and banning wind turbines in the forest). According to its speaker, nearly 100 citizens' initiatives support it and its steering committee consists of about 20 people (Interview, Thomas Jacob). Their two central demands were a general setback of ten times the height of a wind turbine to the nearest house ("10h") and no development of wind turbines in forests. The demand of "10h" originates from the national discussion when federal states were given the right to decide on standardized setbacks for their territory (*Länderöffnungsklausel*). Bavaria endorsed this rule and *Rettet Brandenburg* aimed to prompt their state parliament to follow this example. Setbacks of wind suitability areas are usually decided in the regional planning process and setbacks from particular wind turbines in the approvals process. With a common height of 200 meters, a "10h" setback would usually mean 2,000 meters. The demand of excluding forests particularly targeted the state directive to regional planning which allowed particular kinds of forests to become wind suitability areas after an assessment of the specific site.

In July 2015, *Rettet Brandenburg* submitted 33,000 signatures to state parliament. This was followed by a hearing in September 2015, but their demands were dismissed, prompting the initiators to proceed to a public referendum. From January to July 2016, Brandenburg residents could register their signatures with the local authorities. As only 45,270 signatures were ultimately submitted, the target of 80,000 was not met (MAZ 2016b), despite a mobilization campaign by *Rettet Brandenburg*. The campaign included public roundtable discussions and information stalls all over Brandenburg as well as rallies in the state capital of Potsdam. In order to understand the rationale behind the anti-wind movement in Brandenburg, the following section reflects the viewpoints of *Rettet Brandenburg*'s speaker Thomas Jacob, who was one of the main initiators of both popular initiatives of 2009 and 2014.

Thomas Jacob started to become active in local politics in the early 1990s and has since been mobilizing against the construction of a wastewater system in the area, large-scale poultry breeding farms close to residential buildings as well as the first proposed wind parks in the area. When the first wind turbines were erected in Brandenburg, Thomas Jacob reports that he embraced them as beautiful and a potential replacement technology for Brandenburg's open cast lignite mines to end the involved relocation of peoples' homes and entire villages. Despite this rejection of lignite mining, Thomas Jacob is not entirely convinced of anthropogenic reasons behind the existing "climate catastrophe"[3] (Interview, Thomas Jacob). Outright climate skepticism is indeed common within *Rettet Brandenburg* as well as within the German anti-wind umbrella organization *Vernunftkraft*,[4] of which *Rettet Brandenburg* is a member. At a rally in April 2016 in Potsdam organized by *Rettet Brandenburg* for instance, a representative of *Vernunftkraft* held a speech with many statements directly skeptical of anthropogenic climate change, including the theory that current increases in temperature only result from natural cycles in the earth system, but not from anthropogenic sources (Participant observation in Potsdam).

Thomas Jacob reports to have changed his initial positive stance towards wind turbines upon realization that hundreds of wind turbines were planned in Brandenburg instead of single spread-out turbines and that the rate of lignite mining still increased despite wind power development. At the beginning of wind turbine development in Brandenburg, he states, people lacked experience with turbines. Protest in Brandenburg also increased in the last four or five years because they realized that their concerns were not being listened to. Furthermore, the developments created a negative social atmosphere in villages between those who financially benefitted by renting out their land and others who had to bear the view. A tipping point of protest, Thomas Jacob states, was in 2011 when the state government allowed forests to become wind suitability areas, which led to escalating "public outrage" and anti-wind "citizens' initiatives sprang up like mushrooms" (Interview, Thomas Jacob). As this representation points out, the landscape effects of wind turbines have been a central point of concern. In this context, the speaker of *Rettet Brandenburg* contends that Brandenburg is disproportionally burdened by the *Energiewende*:

> So why should we be the energy state Brandenburg, why should we destroy our [emphasized] beautiful countryside for, seen selfishly, for Baden-Württemberg, Hesse, Rhineland-Palatinate and whoever else that refuses to set up their own wind turbines? (…) So it was like that, Brandenburg – I've always said – is the garbage dump or East Germany is the garbage dump for the Federal Republic. Dutch factory farming facilities that are no longer permitted in Holland get permission to build their huge stables here in Brandenburg with hundreds of thousands of animals. (…) And we [emphasized] should get the whole burden, why? Why are we the Energieland? Just because a few stupid politicians have come up with this slogan, we open our mines so that they can get their coal?
>
> (Interview, Thomas Jacob)

As this quote exemplifies, the question is raised why Brandenburg should bear the burden and sacrifice its beautiful scenery while other German states reject wind turbines. This alludes to the forerunner status of Brandenburg with regard to wind turbine development and the fact that other state governments such as Bavaria have successfully implemented larger setbacks and therefore severely restricted further wind turbine development. The underlying rationale of Brandenburg's ambitious wind power policy – the reference to being an *Energieland* – is defied and contrasted with the view that Brandenburg and eastern Germany in general should not serve as the "waste dump" of the republic, in terms of bearing more burden than other federal states or regions, whether this refers to large-scale livestock farms or wind turbine deployment.

Thomas Jacob's more general and underlying criticism of wind turbine development relates to the prevailing capitalist system. The optimal energy system should not be geared towards maximizing profits but to secure reliable energy supply while "carefully dealing with resources and not destroying the earth in the search for profit" (Interview, Thomas Jacob). In the way the energy system currently works, he continues, humans no longer play a role and that is why wind turbines are not built at least 3,000 meters away from homes or on Brandenburg's remotely situated former military or mining areas. In contrast to former GDR times when the land was public property it now seems to belong to the "lobby" (Interview, Thomas Jacob) that chooses a wind turbine location that maximizes its profits. Nonetheless, he does not wish to see the GDR back, as he is now allowed to openly utter his criticism.

A major point of concern is thus the perceived unequal sharing of benefits and imposed burdens. This is further exemplified by Thomas Jacob's statement that "in principle we don't derive any benefits from the *Energiewende*, electricity gets more expensive, the landscape is destroyed, living conditions deteriorate" (Interview, Thomas Jacob). He continues that this would be different if they got electricity at cheaper prices or for free. This statement relates to the higher electricity prices that indeed have disproportionally risen in Brandenburg compared to other regions in Germany. This is due to the regulation that the costs of developing electricity grids may be added to the electricity prices in those regions where the development occurs.[5] According to Thomas Jacob, democracy is about the will of the majority, but the wind industry, which acquires huge profits, comprises only a small number as compared to those who suffer from it. In the same vein, the "destruction of the landscape" (Interview, Thomas Jacob) only benefits a few companies and is thus similar to lignite mining.

Rettet Brandenburg has engaged in a number of activities such as writing letters or meeting with politicians to garner support from political parties. The group wrote letters to all members of state parliament and delivered them in person to the president of state parliament. Yet, Thomas Jacob reports that those letters were either not answered at all or the group received "some flimsy explanations" in reply (Interview, Thomas Jacob), which would be, in his view, a sign for the glaring discrepancy between the government and the people. There were several occasions on which the group met with officials from the federal or federal state level. Similar to the results of letter writing however, Thomas Jacob depicts these meetings as unsatisfactory, contributing to his impression that politicians are not interested at all in the citizens' proposals or involvement. This once again highlights that the current procedure of wind turbine siting is not perceived as a just and inclusive process. Instead, wind turbine development is depicted as a top-down approach with no benefits for the local population.

Ambiguous political support by the CDU

At the time of the research for this book, the Christian Democrats (CDU) were the most important opposition party in Brandenburg in terms of seats in parliament (21 seats). While the smaller green party (Bündnis 90/Die Grünen, six seats) did not support the anti-wind movement due to the party's positive attitude towards the *Energiewende*, the other two smaller opposition parties of BVB/Freie Wähler (three seats) and AfD (11 seats) fully supported the anti-wind movement by adopting their positions in parliament discussions and giving speeches at their rallies. The AfD represented the second largest opposition party in parliament but holds a particular position as a climate skeptic populist party that has mainly flourished on the narrative that decisions are taken by the political elite over peoples' heads. This viewpoint is premised on their interpretation of the 2015 decision by the German government to open the border to refugees. As the AfD in Brandenburg has been a single-issue party based on opposing migration and has made few efforts to distance themselves from extreme right-wing positions among their members, none of the established parties has been willing to engage in cooperation with them. The AfD hence has no options for governing in coalition with another party and remains marginalized. In contrast, the CDU, as the largest opposition party in Brandenburg state parliament, would qualify as a powerful ally to *Rettet Brandenburg* due to its political force. In order to depict the scope of cooperation the CDU offers to the anti-wind movement, this section first introduces the supportive actions of the CDU parliamentary group and then looks beyond and presents the viewpoint of the CDU spokesperson for energy in state parliament, Dierk Homeyer.

When the Brandenburg state parliament was legally required to address the concerns of *Rettet Brandenburg* because they had submitted enough signatures in support of their popular initiative in July 2015, the CDU parliamentary group pledged to respond positively. This decision followed several parliamentary questions and resolution proposals the CDU group had submitted in favor of applying the *Länderöffnungsklausel* in Brandenburg (Landtag Brandenburg 2014a, 2015c). Supporting the demands of *Rettet Brandenburg* was motivated by the view that Brandenburg had already contributed to the *Energiewende* to a considerable degree and that no further expansion of wind energy to the detriment of human health and wildlife should occur. The adoption of the demanded standardized setback ("10h") would allow "effective health and nature protection without unduly constraining wind energy as a sustainable energy source" (Landtag Brandenburg 2015b, p. 1). In the same vein, the CDU parliamentary group supported the requested rule to exclude forested areas from wind turbine development by emphasizing the role of Brandenburg's forest for recreation, the climate and nature protection.

After the Brandenburg parliament decided to dismiss the demands of *Rettet Brandenburg*, the CDU parliamentary group still demanded changes in Brandenburg's approach to wind energy. In a May 2016 request to state parliament, the group expressed the view that Brandenburg should remain an *Energieland* but needed to tackle issues in social acceptance (Landtag Brandenburg 2016). To this end, the revised version of the Energy Strategy should reconsider Brandenburg's deployment target of wind energy and the goal of devising two percent of its landmass as wind suitability areas. Furthermore, state forest law should be changed to forbid placing wind turbines in forests and a general setback of 1,500 meters should be adopted (Landtag Brandenburg 2016).

As these initiatives show, the CDU parliamentary group has basically backed the concerns of the anti-wind movement in Brandenburg. Yet, as presented earlier, the speaker of *Rettet Brandenburg* reports that they did not feel adequately supported by any political party, including the CDU. In order to elucidate the CDU's view on the movement, the following section draws on the perspective of the spokesperson for energy issues of the CDU parliamentary group (Interview, Dierk Homeyer). Dierk Homeyer states that political forces pushed heavily for the development of wind turbines, which led to substantial deployment rates within a relatively short period of time. Grid expansion was not able to keep up, which led to wind turbine stand-still in times of much wind. This fact together with the effects on the landscape and higher network charges in Brandenburg has led to decreasing social acceptance in Brandenburg:

Let me just say here in Brandenburg the SPD and the Greens, they really whitewashed things. And now it hits them with full force. With full force. The people are against it, they don't want it, they get angry about it and they create citizens' initiatives. And all these initiatives, which cannot really understand it, well they turn to someone else in the case of protest. We are now dealing with all of this. And that is why it is now time to take clever countermeasures.

(Interview, Dierk Homeyer)

As this quote exemplifies, the CDU energy spokesperson especially blames the SPD and Bündnis 90/Die Grünen for having sugarcoated the situation. He argues that citizens' initiatives emerged because people would be frustrated with wind turbine development and "in the case of protest", people would orient themselves "somewhere else". This alludes to the rising significance of the AfD and their recent gains in several German state elections.[6] Homeyer regards this as a driver to become active to prevent them from gaining more power. In this regard, he raises the topic of democracy and communication with anti-wind groups. He believes that it is necessary for both sides to find their way back to "forms of dialogue" (Interview, Dierk Homeyer) in order to treat each other decently and develop sympathy for the other's perspective. Homeyer believes that a proper communication strategy over wind turbine development was missing from the beginning, similarly to the influx of thousands of refugees to Germany in 2015 which led to the rise of the AfD in the first place.

Dierk Homeyer specifies the following measures for improving social acceptance of wind turbines. First of all, it should be conveyed to the people that Brandenburg has achieved enough in wind power development and that further expansion would be considerably limited. This shall be realized by reducing the state deployment targets, excluding forests from wind turbine development, directing the RPGs to establish larger setbacks and working with the federal level to change the Federal Building Code that classifies wind turbines as "privileged facility". Furthermore, Brandenburg's network charges should be decreased to the height prevailing at the federal level and grid expansion should be financially incentivized. Despite this general support for the demands of the anti-wind movement and the perspective that Brandenburg's approach to wind energy needs considerable improvement, the contact to anti-wind groups had not been overly warm. Dierk Homeyer explains the fact that the popular initiative had not been successful by its bad timing and the lack of "political lever" (Interview, Dierk Homeyer). The remainder of this section presents the reasons that contributed to the fact that the CDU in Brandenburg did not become a strategic ally to the movement, which could have resulted in more political influence and support.

Dierk Homeyer reflects that he met with many local citizens' initiatives against wind turbines, but many of them were not interested in making compromises and were unwilling to consider the public interest:

If you are dealing with citizens' initiatives, there is always a fixed rule: Assume that if you go there they are going to be the opponents of a cause, rarely the initiative for something. If you go there, assume that you cannot convince them. It is only about a yes or no. A conditional answer is not enough. So, you can't always be for what a citizens' initiative is for, because a politician is committed to the common good and has to balance public welfare interests and find a reasonable way. Because, sometimes in public opinion this is not enough, we have a representative democracy. Everyone can blather on about whatever they want, no question about it, everyone can also want what they want, but not everyone gets to decide for what they want.

(Interview, Dierk Homeyer)

This quote illustrates that Homeyer regards some citizens' initiatives as following their individual interest only, while politics is obliged to guarantee the public interest instead. Homeyer finds that often, the "idea of the public interest, which is to ask what is actually benefitting everyone" (Interview, Dierk Homeyer), is thus not particularly pronounced within citizens' initiatives. In this vein, he also states that wind turbines had never been a major topic in the run-up of state elections. He explains this by his observation that the topic only affects a small fraction of people. Additionally, Homeyer specifies that many wind opponents he met have a general problem to accept anthropogenic climate change, which makes them completely opposed to the *Energiewende*. According to Homeyer, many of them are retirees who bring folders to the meeting full of studies that supposedly show the non-existence of climate change. This sometimes includes conspiracy theories:

Sometimes that goes along with conspiracy theories, right? And that's what you're seeing here. And it's just very difficult, if someone is so firmly convinced that this whole energy transition is complete nonsense and that climate change doesn't even exist, to convince him that we have to change something in this world and that fossil fuels are also finite. And that we must arrive at a reasonable solution that is fair for generations to come. And that we are the generation that is actually driving this now and must do something. That is difficult.

(Interview, Dierk Homeyer)

As this quote exemplifies, Homeyer regards it highly difficult to deal with local opponents that are intrinsically convinced of climate-skeptic theories. At the same time, Homeyer finds it absolutely necessary to act now to change the energy system in the face of limited fossil fuel reserves and the need to combat climate change.

In conclusion, the CDU as the then largest opposition party in Brandenburg has formally supported the demands of *Rettet Brandenburg*, but on the practical level has been reluctant to act as an outspoken ally to the movement. This is because the anti-wind movement is regarded as neglecting the public interest and as only focusing on their private interest. Second, many anti-wind activists are described as climate skeptics who do not seem to acknowledge the need for the *Energiewende* as the main rationale behind the development of renewables.

Government response

The Brandenburg state government has responded to the increase in opposition to wind turbines continuously and in several ways. Already in 2009, it became aware of lacking social acceptance in the context of a major grid expansion project in North-Eastern Brandenburg, the "*Uckermarkleitung*" (Interview, Ralf Christoffers, State Minister for Energy and European Affairs 2009–2014). Around the year 2010 when the Energy Strategy 2030 was in the process of being developed, the government was aware of five anti-wind citizens' initiatives and social acceptance entered the political agenda (Anonymous interview ministry representative). Generally, the government regards Brandenburg's social acceptance challenges in light of its forerunner status in wind energy deployment. Brandenburg was one of the first to experience challenges early-on and was one of the first to raise issues that had not yet been discussed at the federal level, including energy storage development, system integration and social acceptance and participation (Anonymous interview, ministry representative). Several measures to address concerns about wind energy development were already made part of the Energy Strategy 2030. The following section addresses those and further responses by the Brandenburg state government to anti-wind protest. The response generally includes an information campaign, endorsing the regional planning system, addressing the federal level to change the cost structure and attempts to increase citizen's participation.

Information campaign: acceptance measures in the Energy Strategy 2030

While the Energy Strategy 2020 only alludes to the topic of social acceptance by stating that the strategy "must be communicated comprehensively" and by announcing the development of a concept that is "dealing with regional problems in acceptance" (MW 2008, p. 50), the Energy Strategy 2030 more directly tackled the issue of social acceptance. Its main focus is a strategy built on information and participation:

> The Brandenburg state government takes the concerns of the population seriously and will do everything in its power to win approval for its energy policy. It is committed to a transparent information policy and targeted participation of the population. The aim is to increase the benefits of and trust in an efficient energy supply in order to achieve the broadest possible support for the objectives of the Energy Strategy. In order to be able to successfully implement energy policy goals, the measures must be comprehensible to

the population. On the one hand, the complex supra-regional interdependencies of the energy policy of the Federal State of Brandenburg in the national and European context and the resulting federal responsibility must be presented. On the other hand, concrete solutions must be found on the ground to support the decision-making process.

(MWE 2012a, pp. 43–44)

The strategy thus acknowledges the "concerns of the population" and announces that the state government will work intensively on soliciting support for its energy policy, also by explaining the embeddedness of Brandenburg's energy policy in the national and European context. The Energy Strategy 2030 names "creating regional participation opportunities and acceptance" as one of its six overall strategic targets. They are to be realized by the three pillars of "communication and information", "involvement and participation" and "balance of interests and conflict resolution" (MWE 2012a, p. 53). Furthermore, "acceptance and participation" was added as a fourth aspect to the strategy's overarching energy target triangle consisting of "security of supply", "economic efficiency" and "environmental and climate protection".[7] Three concrete measures to "improve the understanding about the state energy policy" were specified. Those include (1) the development of the internet-based information and communication tool *Energie- und Klimaschutzatlas Brandenburg* (energy and climate protection atlas Brandenburg), (2) a further development of instruments and platforms serving the communication of the Energy Strategy's regional implementation, and (3) the development of innovative financing models that allow local residents to financially participate in renewable schemes (MWE 2012b, p. 47).

In order to solicit support for Brandenburg's approach to renewable energy and respond to declining levels of social acceptance, the then Minister Ralf Christoffers traveled through the rural areas on an *"Energietour"* to engage with citizen's initiatives and other residents at the local level. Ralf Christoffers states that previous consultations with the RPGs revealed that they needed support on the political level (Interview, Ralf Christoffers). Christoffers regards the implementation of the *Energiewende* as a "fundamental issue which de facto changes society as a whole" (Interview, Ralf Christoffers). In such important situations of change, it is imperative to support the local level and he had thus always been in favor of engaging in dialogue instead of imposing decisions in a top-down manner. Meeting citizens at the local level, Christoffers states that Brandenburg's energy exportation has been one of the most frequently discussed aspects. He then replied that Brandenburg is embedded in the national economy and cannot only act for itself:

If we begin to carry out the energy transition in Germany in the small states, then it will fail. And we have experienced a process, and this was also a process of recognition for many, also in the discussion processes that we conducted, energy production has shifted from the south to the north. And, of course, there are completely different requirement profiles associated with this in order to create energy security. And the state of Brandenburg has, as a result of but not only, its coal industry, always had the function of an energy-exporting country.

(Interview, Ralf Christoffers)

As this quote exemplifies, Germany's *Energiewende* serves as a major reference to defend Brandenburg's approach to wind turbine development. As the nuclear reactors in southern Germany are eventually phased out, electricity production would shift to the northern parts. The connected argument of Brandenburg being an "energy export state" further contributes to this argumentation. The *Energiewende* would not be successful if everyone was only thinking for themselves, according to Christoffers, and Brandenburg had always had the role of an energy export state. A staff member of the ministry who had been actively involved in the energy tours recalls that they always tried to communicate that we were living in a global economy and he had raised the question of whether we should also produce our own cars (Anonymous interview ministry representative). Generally, the staff member had the impression that via this pro-active approach and providing the possibility to engage in a discussion, people were encouraged to become active in the first place. He reports that the ministry had always tried to be transparent and publish key documents that were relevant for taking their decisions, but engaging in a factual discussion with anti-wind groups was nevertheless "impossible". Because of this experience with open dialogue and providing transparent information, the ministry representative advocates against giving in to the demands of the anti-wind opposition: "If we did '10h' now, we would then start discussing about 15h". Yielding to some demands of the anti-wind movement would therefore never be regarded as sufficient (Anonymous interview, ministry representative).

Endorsing regional planning: reacting to the demands of the Volksinitiative

In 2015, the discussion about standardized Brandenburg-wide setbacks peaked when the popular initiative of *Rettet Brandenburg* submitted 33,000 signatures to state parliament that demanded the introduction of a general setback of ten times the height ("10h") and no wind turbines in forests. Until the end of the year 2015, Brandenburg formally had the chance to introduce a respective federal state law in the context of the *Länderöffnungsklausel*. The Brandenburg state parliament however rejected these demands by referring to the existing regional planning approach that would best be suited to find locally adapted setbacks (Landtag Brandenburg 2015a). This endorsement of the regional planning approach to find suitable locations and determine setbacks follows a common political line in Brandenburg which is also reflected in the coalition agreement and in SPD policy papers (SPD Brandenburg 2015; SPD Brandenburg and DIE LINKE Brandenburg 2014). At the 2015 *Brandenburger Energietag* (Brandenburg Energy Day), an annual public conference on energy issues in Brandenburg, the Minister for Economic Affairs and Energy Albrecht Gerber reported in a speech of being approached by local residents concerned of further wind energy development in their area. He had replied that he was not able to help them because it was all decided by the RPGs and in the regional assemblies (17th *Energietag* in Cottbus, 03.09.2015). Reference to the regional planning process is therefore not only a common line in policy documents, but also serves as a direct response strategy to deal with protestors.

While state parliament dismissed the two concerns of *Rettet Brandenburg*, the 2015 resolution statement still contained a number of measures to address the organization's concerns (Landtag Brandenburg 2015a). In the context of the coming revision of the Energy Strategy, the state parliament called upon the state government to examine whether the renewable energy targets could be met with an altered area target of wind energy,[8] develop instruments to support local energy concepts to improve citizen participation, and conduct the review of the Energy Strategy in a transparent way. Furthermore, the RPGs should be further supported in their developing regional plans. Concerning the demand of excluding forests from wind turbine development, forestry-related suggestions and expert opinion was proposed to be included in forest mapping and in recommendations to the RPGs. In order to improve the participation possibilities of municipalities with less than 10,000 inhabitants, which are often not represented in the regional assemblies, a draft law should be tabled to strengthen their participation rights in regional planning (Landtag Brandenburg 2015a).

Addressing costs at the federal level: EEG reform and network charge

When Brandenburg fell from first to fifth place in the Germany-wide ranking for renewable energy efforts in December 2014, Albrecht Gerber released an official statement that declining levels in social acceptance were also attributable to an increase in electricity prices for local residents because of Brandenburg's advanced stage of renewable energy development (MWE 2014). As the regulation of this network charge falls under federal jurisdiction, the Ministry of Economic Affairs and Energy demands a system of "fair burden sharing" from the federal government (MWE 2014).[9] As this statement exemplifies, the issue of higher electricity costs in Brandenburg as compared to other federal states is regarded as one of the main reasons behind a decline in social acceptance.

The call towards the federal level to reform the network charges has specifically been raised in the context of the EEG review. In May 2016, Brandenburg's state premier Dietmar Woidke threatened at a meeting of all German state premiers that Brandenburg would reconsider its renewable energy targets if no fairer system of cost sharing is implemented (MAZ 2016c). At the annual *Windbranchentag Berlin-Brandenburg* (wind energy day Berlin-Brandenburg), an annual public conference by the wind industry with invited talks from politicians, the State Secretary of the Brandenburg Ministry of Economic Affairs and

Energy Hendrik Fischer principally endorsed the reform of the EEG to replace the FIT with a bidding system. Steering wind energy development via the FIT had failed, according to Fischer, as grid expansion and the development of energy storage had proven not to be able to keep pace (speech 26.05.2016, Potsdam). In the same vein, the spokesperson for energy policy of the SPD group in state parliament Ralf Holzschuher praises the new EEG as an "important impulse" for the *Energieland* Brandenburg that could improve the acceptance of municipalities and citizens (MAZ 2016d).

Increasing participation: limited room of maneuver

In July 2015, the SPD group in state parliament issued a resolution paper named "ensuring acceptance of wind energy" (*Akzeptanz der Windenergie sichern*). Next to endorsing regional planning and the better involvement of smaller municipalities as depicted earlier, the coming new Energy Strategy should reconsider the two percent land area target. The nightly lighting of wind turbines should only be switched on with approaching aircrafts, grid expansion should be better synchronized and the development of energy storage should be supported. Furthermore, financial participation of surrounding municipalities shall be regulated by law and a fund to support measures of municipalities to improve acceptance should be implemented (SPD Brandenburg 2015). As much as these proposed measures sound like a well-conceived agenda to improve social acceptance, they are regarded as highly difficult to implement in practice (Interview, Ralf Holzschuher). According to Ralf Holzschuher, the room of maneuver for Brandenburg to implement measures for a better financial participation is limited. For instance, the model of the wind investor funding a sports ground is often not legal because it might be taken as a form of bribery, according to Holzschuher. He is also skeptical about the law that Mecklenburg-West Pomerania has passed to urge wind developers to offer a share of the wind project to people living in the vicinity, as this law intervenes with private property rights and may surpass Brandenburg's legislation authority (Interview, Ralf Holzschuher). Overall, finding appropriate solutions to have Brandenburg's residents better financially participate in wind energy development is regarded as highly restricted.

As a result, Brandenburg did not support measures for improving financial participation of citizens or municipalities in wind turbine development at the time of the research.[10] A ministry representative states that there have been some minor changes in tax laws that made it easier for municipalities to reap benefits from wind turbines (Anonymous interview ministry representative). This was the result of a study the government commissioned in 2012 to explore potential models for citizens participation. Nonetheless, the ministry representative does not consider it primarily Brandenburg's responsibility to secure increased participation of municipalities and citizens. He refers to the new EEG (2017) that would make it difficult for small-scale community projects to compete against larger companies anyway. He also hints to the consultation process for the Energy Strategy 2030 where they discussed possible measures to increase participation. The ministry representative reports that during these meetings, people often joined the workshops and ranted instead of providing concrete suggestions of how to improve: "We do not sit here in an ivory tower and have exclusive knowledge (...) therefore, a lot has to come from the base, too, I'd say" (Anonymous interview, ministry representative). Some of the suggestions were not constructive, such as ending wind energy altogether. They still collected a large pool of proposals, but many were impossible to implement due to legal concerns (Anonymous interview, ministry representative).

In summary, the Brandenburg government has taken the anti-wind movement seriously by addressing their demands in several ways. An extensive information campaign was intended to garner support for Brandenburg's renewable energy policy and its contribution to the *Energiewende*. The government, however, chose not to take over the two specific demands of *Rettet Brandenburg*, although they had the room of maneuver to implement those changes. Instead, the state government referred to the existing regional planning approach to best be able to address them. In a similar vein, the state government appealed to the federal level to address the cost question, which the state government identified as a core reason behind the opposition. A fourth way of reacting to anti-wind protest was to elaborate on new measures to increase the financial participation of local communities, but the government had difficulties in identifying possible solutions at the time.

The local level: a case study of RPG Prignitz-Oberhavel

The remainder of the chapter zooms into a local case study of anti-wind opposition in Brandenburg. The purpose of this local focus is to shed light on the origins and characteristics of local citizens' initiatives, their view on wind turbine development and their channels of influence. The focus is on the RPG Prignitz-Oberhavel, which has a forerunner status among Brandenburg's RPGs in terms of its installed wind power capacity.

The RPG Prignitz-Oberhavel is situated in the north-western part of Brandenburg and spreads over 6,428 per km². Encompassing 68 municipalities, it comprises the three *Landkreise* of Prignitz, Ostprignitz-Ruppin and Oberhavel. In 2012, the RPG had 380,100 inhabitants and a population density of 59 inhabitants per km² (GL Berlin-Brandenburg 2016). Due to its vicinity to the capital city of Berlin, the *Landkreis* of Oberhavel is more densely populated, while Ostprignitz-Ruppin is less densely populated and the far north-western *Landkreis* of Prignitz has the lowest population. The regional assembly of Prignitz-Oberhavel consists of 32 regional councilors out of which half are elected and half comprise the *Landkreis* commissioners and mayors of municipalities with more than 10,000 inhabitants (Kuschel 7/8/2015).

Prignitz-Oberhavel is a forerunner in renewable energy development and is depicted in its regional energy concept as a sustainable and active energy region (*nachhaltige und aktive Energieregion*) (RPG Prignitz-Oberhavel 2010). Its installed renewable capacity is 1.3 GW, which represents about 24 percent of the total installed renewable capacity in Brandenburg (MWE 2016). In 2013, the majority of Brandenburg's then existing 3,229 wind turbines were situated in RPG Prignitz-Oberhavel (936), followed by RPG Lausitz-Spreewald (652), RPG Uckermark-Barnim (640), RPG Havelland-Fläming (606) and RPG Oderland-Spree (396) (Landtag Brandenburg 2014b). In June 2016, Prignitz-Oberhavel still came in first in Brandenburg with 953 wind turbines (Interview, Ansgar Kuschel). More than 520 turbines are situated in the *Landkreis* of Prignitz, more than 330 in Ostprignitz-Ruppin and more than 70 in Oberhavel. Due to its higher population number, Oberhavel is represented by more councilors in the regional assembly than Prignitz. Among the elected regional councilors, nine are from the *Landkreis* of Oberhavel, four from Ostprignitz-Ruppin and three from Prignitz (Kuschel 7/8/2015).

Prignitz-Oberhavel has experienced strong controversies over wind energy deployment at the regional planning level. The RPG has been in the process of developing its regional plan for more than ten years and in 2017, the process is still ongoing. In 2007, the regional assembly decided to start the process of devising a new regional plan to replace the outdated 2003 version. The first new draft was presented in April 2015. The legally required public consultation process from June to August 2015 yielded more than 2,000 comments. Out of the 35 proposed wind suitability areas, two were exclusively confirmed, three exclusively rejected, three proposed for enlargement, and 24 were confirmed ("keep" or "enlarge") by some and rejected ("withdraw" or "reduce") by others. Additionally, 80 new suitability areas were proposed (RPG Prignitz-Oberhavel 2016). The municipalities rejected 19 proposed areas and proposed seven further areas to be included. Furthermore, they asked to reduce eight proposed areas and assess whether seven areas could be enlarged (RPG Prignitz-Oberhavel 2016). In response to this strong controversy around the draft proposal of the regional plan, the regional assembly decided in 2016 to prepare a second draft involving a second public consultation process. The regional assembly of Prignitz-Oberhavel adopted this second draft of its plan, called "free spaces and wind energy" (*Freiraum und Windenergie*), in April 2017 and as of July 2017, almost 1,700 statements from the public consultation process are being reviewed (RPG Prignitz-Oberhavel 2017).

Addressing the tense debate around wind developments, which has contributed to the slow process of devising its regional plan, the RPG has openly put forward policy demands and has also explored options to introduce larger setbacks at the regional planning level. At the end of the 2010–2013 process of developing its energy concept, the RPG has presented the following policy demands (Kuschel 7/8/2015, p. 8). First of all, the further procurement of renewable energies should be better aligned with grid expansion and also with social acceptance in general. The costs for grid expansion and operation should be shared among all federal states and the EEG should generally be reviewed with regard to its cost structure. Moreover, those regions producing electricity should financially participate in the energy transition, which would necessitate a nationwide definition of "energy region" and the development of a financial bonus system as incentive and motivation. These demands exemplify the fact that at the regional planning level, burdens and financial benefits are a major point of concern. As the RPG produces more renewable electricity than other regions, the RPG calls for financial compensation and regional balance.

The head regional planner's perspective

Orchestrating the development of regional plans and the protest against the designation of wind suitability areas within the planning process, the head regional planner in Prignitz-Oberhavel possesses a uniquely insightful understanding of the characteristics of anti-wind protest in Prignitz-Oberhavel. Ansgar Kuschel recalls that in the 1990s when wind turbines were first set up in Brandenburg and Prignitz-Oberhavel, a general enthusiasm prevailed in the region and wind energy was regarded as a promising future topic as well as a sustainable and future-proof technology (Interview, Ansgar Kuschel). Many municipalities embraced wind power, media articles were positive, and the erection of new wind turbines was at times celebrated with a wind farm party with music and free beer. Kuschel reports that by the mid-2000s an anti-wind backlash started to become apparent. At that time, people became aware of the dimension of wind energy development and new turbines started to be higher than 100 meters which, by law, required lighting. According to Kuschel, the flickering lighting led to a first public outcry revolving around "the never ending flashing during day and night". This was reinforced by people becoming aware of the ongoing regional planning process, which made them realize that not only single new wind turbines were planned but 45 new wind suitability areas. When the Energy Strategy 2020 released in 2008 proclaimed that Brandenburg is well on its way, but needs to double the number of existing wind turbines, it was regarded as a "declaration of war" (Interview, Ansgar Kuschel).

Next to the sheer numbers of planned wind turbines, the head of the planning office reports that another important point of concern is the public consultation process (Interview, Ansgar Kuschel). Many people objected that the process was inaccurately named, as it would not involve real public participation. Kuschel admits that the planning process is heavily prescribed and that the planners have to abide by the rules and cannot withdraw a proposed area because of particular raised concerns, such as feared property devaluation. The potential for including local input and finding locally adapted solutions has however changed over the course of time. The 2003 regional plan, which did not yet legally require public consultation apart from municipalities and specialized authorities, and a 2010 draft of a new plan that was later abandoned were "plans of compromise". They were largely based on the plans and planning conceptions of municipalities and investors. This considerably changed with the rule of the Higher Administrative Court Berlin-Brandenburg (OVG: "*Oberverwaltungsgericht*") that there must be a specific procedure after which wind areas are designated, precluding the option of just directly taking over the municipal plans. Kuschel therefore evaluates the previous plans and drafts as following a philosophy of finding the most consensus-oriented areas for wind development, while the strict process that the OVG court rule prescribes is more prone to resistance. The new required procedure has been one of the most important sources of criticism, as Kuschel reports. The RPG has often been accused of a dogmatic planning style without compromising. Indeed, Kuschel regards the RPG's room of maneuver as small:

> There's very little room for maneuver. That is also the difficulty for the regional councilors, which have noticed, well, we started here 25 years ago, saying, we will shape the region, only to be told now, yes, you can shape the region under the conditions of 100 judgments, regulations and laws, and please make sure to apply them all and not to your own taste, we as Prignitz or we as Oberhavel are doing this or that now. In other words, there is a very strong disillusionment in the regional councils, we are on a very, very short chain as far as the legal framework is concerned.
>
> (Interview, Ansgar Kuschel)

Kuschel thus reports an increasing disillusionment and disappointment among the regional councilors of the regional assembly. They expected to be able to shape the region's development but now realize that in the face of a myriad of rules and laws, their ability to act is heavily constrained. One possible way to still include local input under the current requirements is the definition of "soft"[11] criteria in the process of designating a wind suitability area. For example, the RPG successfully adopted the soft planning criteria that no village should be encircled by wind turbines by more than 180 degrees. The RPG furthermore introduced several other soft criteria that relate to the size and distance of wind suitability areas. Within the obligation to designate around two percent of the landmass as wind suitability area and all other legal requirements, there is thus still some room to shape wind turbine development and address some local concerns especially with regard to aesthetics.

The local anti-wind storylines (Temnitzregion)

The region of Temnitz, named after a small river, is situated approximately 10 km west of the regional capital town of Neuruppin in the *Amt* Temnitz, which is an amalgamation of six municipalities in the *Landkreis* Ostprignitz-Ruppin (see Table 4.2). Four new wind suitability areas were proposed in the 2015 draft of

Table 4.2 The municipalities of the region of Temnitz

	Surface area in square kilometers	Inhabitants (as of August 2015)	Population density (individual per square kilometer)
Temnitzquell	65.5	759	11.6
Walsleben	31.8	812	25.5
Temnitztal	51.9	1,444	27.8

Source: Amt Temnitz 2016b.

Prignitz Oberhavel's regional plan in the municipalities of Temnitzquell, Walsleben and Temnitztal.

More than 50 citizens joined together in March 2014 to form the citizens' initiative *Keine neuen Windräder in der Temnitz-Region* ("No new wind turbines in the region of Temnitz") (Gegenwind Temnitz 2016a). Their mission statement affirms that the group is not against wind energy or renewable energy in general, but against its "disproportional development to the detriment of humans and the environment" (Gegenwind Temnitz 2016b). Declaring that the existing wind parks already extend over more than one third of Temnitz' surface area, they argue that the district has fulfilled its goals with regard to Brandenburg's ambitions for renewable energy development and thus contributed enough to the *Energiewende* (Gegenwind Temnitz 2016b). The group's website provides place-protective reasons for resisting the further designation of wind suitability areas, referring to the need to prevent their region from turning into an "industrial park" to the detriment of the landscape, nature, stillness and health:

> We are united by the conviction that we must protect our environment and landscape from being transformed into an industrial park for generations – at the expense of our quality of life, historical cultural landscape, natural diversity, recreational silence and health!
>
> (Gegenwind Temnitz 2016b)

The initiative's website problematizes the geographic distance between where electricity is generated and where it is consumed, equating it with the distance between occurring burdens and benefits. While the turbines produce electricity "here", it is consumed in other federal states and abroad but "we have to foot the bill in terms of high electricity costs and the loss in property values" (Gegenwind Temnitz 2016b). In the context of the popular initiative *Rettet Brandenburg*, the group also joined with others to form the larger Action Alliance Headwinds Prignitz-Ostprignitz-Oberhavel (*Aktionsbündnis Gegenwind Prignitz-Ostprignitz-Oberhavel* in November 2014 (Gegenwind Temnitz 2014).

The remainder of this section reflects the storylines of local anti-wind activists. Five aspects will be presented here: their initial opinion and sources of information, their main arguments and view on the energy system, their perspective on the space of participation, their group activities and their relationship with political parties. The interviewees comprise three members of the citizens' initiative *Keine neuen Windräder in der Temnitz-Region* and two communal representatives. One is a local councilor and member of the citizens' initiative, while the other is an honorary mayor and member of the SPD.[12] At the time of the interview, in July 2016, the group reports that the number of people actively involved in their group was approximately ten.

Initial opinion on wind turbines, sources of information and forming opinion

The members of the anti-wind group describe their first contact to wind turbines as rather positively, either because they were not directly affected by them or because they found them "small and cute" (Interview, municipal representative, MR). This is also the reason why they see the first wave of wind turbines in Brandenburg as not having met with strong resistance. As more wind turbines were put up, it became a matter of scale. Whereas wind turbines used to be small and only few in number, their number as well as their appearance is now regarded as overly intrusive. AF reports that the previously small wind turbines were accepted also, "because the negative impacts were unknown at that time, and they did not take these excessive dimensions as nowadays" (Focus group discussion, Brandenburg, AF). The group members TB and FS report that they became active against wind turbines because an existing local anti-wind group distributed flyers and put up posters in public space which made them aware of the risks associated with wind turbines. They consequently attended a meeting of the existing citizen's initiative in the village of Manker to learn more about it, "and yes, we are opposed ever since" (Focus group discussion, Brandenburg, TB):

> Yes, naturally it was all a bit of new territory for us. That means, I was I don't know how many hours underway on the internet in order to grasp the entire magnitude of the catastrophe. And the way it is, one thinks, it can't get any worse, but the further one goes in the material the worse it gets. We went to the meetings, of course, and Action Alliance, too, and so I think that's how one gets started.
>   (Focus group discussion, Brandenburg, anonymous participant, FS)

This quote shows that the previously existing anti-wind group was an important initial source of information for FS and TB, but online research also considerably contributed to forming their opinion. FS reports that the more she got to know about "the issue", the more she became concerned about the "magnitude of the disaster".

Main arguments and view on energy system

The group's main concerns about wind turbines are their impact on the landscape, on health and economic impacts, all of which are embedded in the view that their region has already contributed enough to the *Energiewende*. As the earlier-cited mission statement on the group's website exemplifies, a major criticism is the perception that their landscape is being turned into an "industrial park". According to JO, their area is one of the last open large free spaces in Germany and everyone with a "little attachment to nature" (Interview, JO) should understand that this "silverware" (Focus group discussion,

Brandenburg, AF) must be protected. The fear of destruction of the landscape is accompanied by the concern that wind turbines are harmful to human health. JO is sure that acceptance would increase if there was an end to the insecurity regarding impacts on human health. This insecurity should be addressed by scientific studies. Yet, the interviewees articulate frustration that the government does not sufficiently consider this aspect. The feeling of not being heard and the population being under a health threat contributes to the feeling of injustice:

> So I think with us it is also just this injustice or this, how should one say? It simply goes against one's own sense of justice, that one says this is not fair. Yes? And why does nobody listen and why are there countries like Poland, Denmark, Canada, Australia? They say, "Stop, the health of our citizens is important, we have to take a look here, something isn't right here". Why does Germany say, as so often in many topics, "Well, we don't have any studies yet. But we are not requesting any either". And then the problem does not exist. And there you really feel a little bit . . .
>
> (Focus group discussion, Brandenburg, anonymous participant, FS)

This quote exemplifies that the group feels wronged because their health concerns are not taken seriously. While other countries would commission studies to investigate those concerns, Germany denies the problem by referring to the lack of studies but at the same time not commissioning any studies. The state authorities are also described as denying the issue. AF reports of her complaint to the State Agency for the Environment (*Landesamt für Umwelt*) about the noise levels of existing wind turbines in her neighborhood but she reports having received the reply that she might be overly sensitive. AF reports that when she came back from work on the day the turbines in her neighborhood started operation in December 2014, she was very surprised by a droning noise comparable to an aircraft. Her first thought was to sell the house. Her bedroom faces the West and in times of strong wind she does not feel like opening the windows because of the noise:

> And because of that no fresh air, because of that unhealthy sleep, because of that poor concentration. And if I really get into it and it doesn't go too well for me, I'm afraid of not doing my job properly either. Then I can't even earn a living.
>
> (Focus group discussion, Brandenburg, anonymous participant AF)

This quote illustrates that the noise produced by the wind turbines is regarded as a severe threat to the health and economic well-being of those living close-by. Even if AF states that she is more concerned for her health than about the economic part, those two aspects are in fact related, as AF worries about not being able to work anymore when she suffers from sleep disturbance caused by wind turbine noise. This is further exemplified by the following quote:

FS: But this continuous and steady, this goes – this is like torture, it's a torture method.

AF: Not steady, impulsive. This "schwumm-schwumm", this droning.

FS: Yes, I mean it always stays, you know? A truck comes and drives by and off he goes.

AF: And yes, and there is nothing one can do about it. Right? (Focus group discussion, Brandenburg, anonymous participants FS and AF)

Wind turbine noise is described as a method of torture because it represents a continuous and recurring sound not comparable to the sound of a passing truck, which would disappear after a while. These concerns about health impacts are further exacerbated by the interviewees' strong feelings of injustice that more wind suitability areas are proposed to the Temnitz region, despite the fact that other areas still have far fewer. According to JO, their region has already exceeded the two percent target of Brandenburg's Energy Strategy because it has 2.8 percent of its surface designated as a wind suitability area (Interview, JO). In his opinion, there should be no further wind turbine planning in their region as they already made their contribution to the *Energiewende*:

> And we have already achieved that goal. Right? And I think to myself, it's time to pull the emergency break. To make it fair, it would make sense to shift the missing percentage elsewhere. There is enough surface.
>
> (Interview, JO)

As the target is already met, there should now be an "emergency brake" to halt further wind turbine development. The missing capacity should be procured in other areas or federal states such as Bavaria. This is connected to their general conviction that the Temnitz area, Prignitz-Oberhavel and Brandenburg as a whole have done their part and now it should be the other's turn. This also relates to the concern that the rural areas will be turned into industrial landscapes in order to provide electricity for the urban areas. JO calls Berlin a "greedy behemoth" (Interview, JO) and states that he understands their need for electricity but then they should also adequately pay for it. In the same vein, MR posits:

> And what I have a problem with is, I'll say, I'll also say, the saving, we save here in the country – we make everything available: We provide the space, the noise pollution we have to endure, we have no returns. And Berlin, what does Berlin do for us? Zero.
>
> (Interview, municipal representative MR)

This statement shows that MR regards it as intrinsically unjust that the rural areas provide the land and bear the noise pollution while not financially participating in wind turbine development. At the same time, the capital of Berlin does not do anything in return. Overall, there is no sympathy for Brandenburg's energy policy:

We have now addressed so many points where it is always going against the people. And it is really this state policy that we are talking about. Because one wants to implement this federal policy so obediently. And I don't understand why the state of Brandenburg has to produce so much electricity – I know, 2.3 times what we actually need – why do we have to make a name for ourselves at the expense of our health? At the expense of our silverware, our landscape, and so on. I find it so insane, so absurd. I can't deal with it.

(Focus group discussion, Brandenburg, anonymous participant AF)

The quote also hints at the feeling of injustice that their area produces much more electricity than they need themselves, to the detriment of residents' health, the landscape as their "silverware" and other aspects. This perspective of already producing enough electricity for the region's own consumption is embedded in the view of large-scale wind farms as a "centralized" type of energy production. In contrast, "decentralized" in their perspective relates to energy generation for local energy use. In this vein, JO is convinced that "we have enough electricity for our households – we don't need any further electricity" (Interview, JO).

The interviewees also raise concerns about the economic impacts of wind turbine development in their area, although this issue is raised to a lesser extent than the landscape and health concerns. The mission statement on their website cites high energy prices and a loss in property values among their important arguments against wind turbines (Gegenwind Temnitz 2016b). The municipal representative MR is convinced that wind developers only make profit because of subsidies and the network charges on the citizens' electricity bill. In 20 years when the contracts will have expired, MR is convinced, the area will be visually characterized by "investment ruins" (Interview, municipal representative MR). FS also criticizes the fact that network charges are high in Brandenburg because of the large number of wind turbines, but "we have nothing to gain from it" (Focus group discussion, Brandenburg, FS). Furthermore, FS is convinced that the lack of existing power storage will lead to interruptions in turbine operation and the overproduction of power that will not be used, while the common citizen still has to pay for it via their electricity bill. Yet, health concerns are more important to AF. If she were to be offered a discount on her electricity bill, she states that this would not buy back her health.

The *Energiewende* in general is regarded as a distribution of burdens. Speaking ironically, MR specifies that if it were something positive to strive for, all municipalities would be happy to be allotted wind suitability areas. As the *Landkreis* Oberhavel has far fewer wind suitability areas than the other two *Landkreise* of RPG Prignitz-Oberhavel, there should have been a public outcry in Oberhavel for not being allowed to sufficiently participate in the *Energiewende*. However, the group asserts that they want the nuclear phase-out and that they are not general opponents of wind turbines. The way in which the energy transition in Germany has been done however, is regarded as too fast and the consequences for human health should also be better taken account of. For TB, it is evident that wind and solar cannot realistically be stopped anymore, but the development in Brandenburg has just been too fast. There should be better solutions for power storage and wind turbine planning should also be improved.

*View on space of participation (regional planning): no real
participation*

Not surprisingly, the interviewees are disappointed about the regional planning process in Prignitz-Oberhavel. The common citizen would be overwhelmed by the whole process and would not understand the complicated procedure. AF heard the word "regional assembly" for the first time at a local public meeting on the planned wind suitability areas in the Temnitz region. Being asked how the regional planning procedure of Prignitz-Oberhavel went so far, FS responds:

> Well, first you have to know that such a thing exists. Second, because you have internet. And third, if you actually understand it, and to do that you have to have a lot of knowledge about the whole thing. Then you would have a chance. Period. Everyone else is simply out. That's not participation, certainly not public and not by the citizenry. In my opinion.
> (Focus group discussion, Brandenburg, anonymous participant FS)

The regional planning process is thus not regarded as "real" citizen participation. People would only have a chance to participate in the process if they were, first of all, aware of the very existence of the procedure. They would furthermore need a lot of knowledge to understand the workings of the process. The regional planning process is regarded as utterly complicated and a "complete mess" (Focus group discussion, Brandenburg, AF) and the meetings of the regional assembly as a "stress program" (Focus group discussion, Brandenburg, AF). The regional planning process would not adequately take into consideration the topics that are important for the group, for example the health issue:

> When Mr. Kuschel goes on to say "Yes, the infrasound can only be measured up to a maximum distance of 700 meters, it cannot be measured any further, so therefore it does not exist. There are no health studies, so there is no evidence of any harmful effects, so we don't need to consider anything, you can plan further".
> (Focus group discussion, Brandenburg, AF)

This quote shows again that the interviewees do not feel taken seriously by the authorities regarding their health concerns. They accuse the head of the regional planning office of an attitude of denial: If there are no health studies on infrasound, there can be no impacts. This concern is also linked to their perception of unfairness that in order to have a real say in the process, one must provide opinions from expensive experts. This for example relates to health studies or to environmental expert reports showing that a particular area shall be excluded from wind turbine development because a protected species is present.

The fact that the development of a new regional plan in Prignitz-Oberhavel has taken several years is regarded as outright malice and strategy on the part of the authorities. TB states that, "this is tactics, they want to wear us down" (Focus group discussion, Brandenburg, TB). Yet, the interviewees believe that if they do not act on it now and prevent further wind energy development, the erection of wind turbines will never cease and the next two percent area target will come for sure. Another important concern is the composition of the regional assembly. The Landkreis Oberhavel with the least wind suitability areas has the largest number of councilors in the regional assembly because it has the largest number of inhabitants. According to JO, this is undemocratic because Oberhavel has the majority and can easily dominate decisions in the regional assembly. Instead, representation in the Regional Assembly should be distributed according to land surface area, because this would better account for the number of potential wind turbines. Furthermore, only municipalities with more than 10,000 inhabitants are represented by a councillor in the Regional Assembly. According to MR this devaluates the weight of a rural inhabitant's voice in favor of a more urban inhabitant. Furthermore, the staff of the RPG's planning office is not regarded as independent, because they are directly employed and directed by the state government.

The interviewed municipal representatives regard the regional planning process as serving a "purely industrial interest" (Interview, MR). They generally regard the development of wind turbines as a singular or private interest. By contrast, the municipal representatives see themselves as representing the public interest. They regard it as their duty to fight for the municipality's well-being, or else wind turbines will be forced upon them without any benefit for their community. Only those who rent out land and the wind energy industry benefit from wind turbines. They feel that their municipalities were informed rather than really involved in the decision-making.

Summing up, the group finds the regional planning process very difficult to understand for the lay citizen. The required procedure to submit comments during the public consultation process is too complicated and hinders the ability of more people to participate. This is why one of the core activities of the group is to spread information on the regional planning process and to assist people in submitting their concerns.

Group activities: spreading the word

The group has been undertaking various activities such as collecting signatures for the popular initiative *Rettet Brandenburg*, mobilizing for the public referendum, letter writing to politicians, joining rallies in the capital of Potsdam, and informing people about the regional planning process:

First and foremost, we set ourselves the goal of enlightening others. Because most people, the citizens, they don't know anything about it.
(Focus group discussion, Brandenburg, anonymous participant TB)

As the regional planning process is regarded as utterly incomprehensible for the people, sharing information about the ongoing process is thus one of the group's core activities. This involves sharing leaflets, putting up instructions on how to participate in the public consultation process in the village announcement boards and talking to neighbors directly. While the members of the group also try to attend the meetings of the Regional Assembly if they can, the communal representative MR reports to often joining them to ask questions as a guest. She has already been successful in delaying the decision-making process of the assembly by pointing to the lack of an environmental report, which the procedure required before proceeding to the next step. Moreover, the group organized several panel discussions with politicians and information events at which they shared information about the planning process and also about health issues. AF reports that they once invited experts from Berlin who gave a talk about the impacts of infrasound which "really shocked" (Focus group discussion, Brandenburg, AF) the attending residents.

Despite their efforts, the representatives of the citizens' initiative find it difficult to mobilize more people to join their struggle. When they went door to door to collect signatures for the popular initiative, they were successful in having people sign up, although some of their neighbors declined, asking whether they wanted to have nuclear reactors instead. FS assumes that most people may have only signed to do them a favor. Yet, the group could "truly" (Focus group discussion, Brandenburg, FS) convince some of them, but no-one joined their group as a result. Similarly, when the group held a public meeting to assist people in writing formal complaints in the context of the public consultation process, only a few local people showed up. Nonetheless, the interviewees firmly believe that the majority is actually against wind turbines, but just remains inactive. Those people might sign a petition, but they would not get active in their group. According to the interviewees, this is because people are "lazy" in general, but also due to the fact that they are unaware of the downsides of wind turbines:

Because the information isn't there, the people still think that wind power is a really great thing. But they have no idea about all the negative aspects.
(Focus group discussion, Brandenburg, anonymous participant TB)

The impression that people "still believe" that wind turbines are a good idea endorses the group's emphasis on spreading information through information events, leaflets or their website. Next to the absence of knowledge, the interviewees believe that people are generally not interested in aspects of environmental protection or noise issues. TB believes that because people now notice the increase in their electricity bills, this might offer a better opportunity for mobilization. Another explanation for a weak local support is identified in the persistent mentality leftover from the former GDR:

So it's really like that, someone had to explain to me in Berlin – so with such a company there are of course many, I'll say, from the former GDR. And there they first explained to me what this GDR factor is, the behavior. The behavior is the following: Why should I involve myself in a cause? I've learned long enough, no matter what I do, I don't change anything. And there's something to that. I don't want that to be true, but it often appears to still be the case.

(Focus group discussion, Brandenburg, FS)

The group's observation that many people remain inactive is therefore also explained by people's experience that one cannot effect political change anyway. This is exacerbated by the fact that the fight against wind turbine development has been going on for a while but has not produced visible results. AF reports that a neighbor asked her why she invested so much time, arguing that because there was big money behind wind turbines and, "you will not be able to prevent it anyway" (Focus group discussion, Brandenburg, AF). The reason for not mobilizing more local support is therefore explained by the absence of information, people's laziness and the negative experience of political participation not being able to effect change.

View on and interaction with politics

Many of the group's activities focus on the local level. The municipal planning endeavors of the *Amt* Temnitz (the amalgamation of six municipalities) are thus also at the focus of the group's attention. In 2015, the *Amt* Temnitz announced the development of a municipal plan and proposed several wind designation areas in their area. While the intention of this municipal planning was to restrict wind turbine planning outside these municipal designation areas and thus to actively shape wind power development in their jurisdiction (Amt Temnitz 2016a), the fact that the municipality engaged in planning at all was seen in a negative light by the group. AF criticizes that the *Amt* Temnitz still prepares a municipal plan although they already have a number of wind suitability areas that far exceed the two percent area target. Although all six municipal respresentatives of the *Amt* Temnitz would be in favor of the "10h" rule, their joint municipal planning did not implement this rule because they fear being sued, according to the group:

FS: Fear of preventative actions, yes.
AF: Yes, and this fear and this pressure, because there is so much money. They can afford the best lawyers and, and, and, and yeah, that's really hard.

(Focus group discussion, Brandenburg, anonymous participants FS and AF).

This assumed fear of being sued refers to the multiple court verdicts that influenced the regional planning process. Furthermore, the wind energy industry is regarded as another source of pressure upon the municipality because it can afford the best lawyers to engage in legal cases.

Next to the municipal level, the Brandenburg state level is also regarded as letting the group down. JO considers the Brandenburg political level as arrogant as it does not act on their concerns and generally neglects the rural areas. According to him, the Energy Strategy 2030 exemplifies that there is no strategy for the rural areas:

> There is not all that much content for us, for the rural area. Except that they want our land. Right? And aside from that the SPD has made a decisive mistake, saying "The village idiots, demographically there is only the bacon belt, that's who we'll support. And we'll let the village idiots die out".
>
> (Interview, JO)

This quote underscores the view that wind turbine development is implemented at the expense of the rural areas. While the affluent suburbs of Berlin are promoted, the villagers are doomed to die out. The only thing that would be of interest is the rural area's landmass. JO, who is a member of the SPD himself, believes that this is the last time that the coalition between SPD and DIE LINKE will exist in Brandenburg, because he expects them to make substantial losses in the 2019 state elections. Calling himself a "democrat" (Interview, JO), he is convinced that the AfD will gain substantially by mobilizing dissatisfied non-voters. Furthermore, the municipal representatives express unhappiness about how they were treated during the process of the popular initiative. MR describes that they are in good contact with their representatives in state parliament, but in the end, those would not act accordingly:

> So I'd say we have good contacts with our state parliamentarians, we have conversations. But when it comes to elections and hand signals afterwards, all of a sudden there is a completely different hand on top or none at all where we would have expected it to be, they should have – as it was presented to us during discussions – gotten our approval. Why is the handshake missing?
>
> (Interview, municipal representative MR)

While the general relationship to their MPPs is thus described as rather positive, there is disappointment that in the end, their conversations would not lead to any change in the MPP's voting behavior. Summing up, the interviewees do not only feel "let down by the state government" (Focus group discussion, Brandenburg, AF), but also by the municipal level:

> Yeah, one feels like Don Quixote, that's how it is. And the windmills stand in Potsdam. So in Brandenburg one feels pretty left alone. From politics. From the state government and also at times of course from the municipal government. It is difficult.
>
> (Focus group discussion, Brandenburg, anonymous participant FS)

This quote summarizes the deeply felt feeling of being left alone by politics and of fighting "wind mills in Potsdam", comparable to Don Quixote. As this section has shown, the local anti-wind groups and the interviewed municipal representatives of the Temnitz region are disappointed by the municipal possibilities to help them, the participation opportunities of the regional planning process, and also by the Brandenburg state level.

Notes

1 This was observed by the author in the RPG of Prignitz-Oberhavel on the April 2015 and June 2016 regional assembly and the May 2016 meeting of the RPG's Planning Board.
2 Citizens initiatives are defined here as a group of citizens that share a common understanding about wind turbines in their area and engage in protest. They include registered legal associations (*"Verein"*) but also a group of neighbors without official status.
3 Note: all forthcoming quotes were translated from German into English.
4 "Vernunft" means reason, rationality, sanity; and "Kraft" means strength and (also electric) power. The term is an allusion to electricity generation, e.g. the German term for wind power is "Wind*kraft*". Vernunftkraft mobilizes against the German *Energiewende* (Energy transition) throughout the country.
5 Network charges increase in areas with high electricity grid expansions. This translates into higher electricity bills for consumers who live in areas with high levels of renewable energies. Due to the high deployment rate of renewables in Brandenburg, households in Brandenburg have the highest network charges of all federal states. An average household in Brandenburg is charged 325 Euro yearly, while in Baden-Württemberg, it is 100 Euro less (MAZ 2016c). Furthermore, if the grid cannot feed in all renewable energy generated (on windy days for example), the wind turbine operator is still paid (by 95%) and this is again fed in to the network charges.
6 In September 2017, they were also elected to the German *Bundestag*.
7 The advisory committee for sustainable development of Brandenburg (NHB) positively recognized the introduction of the fourth pillar, but criticized that its plans for implementation remain vague and that the Energy Strategy 2030 had not been devised by means of a broad societal discussion, but instead only by restricted participation of selected representatives from academia, economics, politics and associations (NHB 2012).
8 Reaching the same target with a smaller land surface is possible by the replacement of older, less efficient wind turbines by higher wind turbines with stronger capacity ("repowering").
9 Despite these obstacles for renewable energy deployment in Brandenburg, the ministry regards the fifth place acquired in the 2014 comparison between federal states as an endorsement for the achievements of Brandenburg's renewable energy sector (MWE 2014).
10 In June 2019, the Brandenburg state government passed the *Windenergieanlagenabgabengesetz* (BbgWindAbgG) (wind turbine levy act) which prescribes the operator of each wind turbine becoming operational from 1st of January 2020 to pay an annual levy of 10,000 Euro to the surrounding municipalities ((Landtag Brandenburg 2019)).
11 "Soft" criteria, in contrast to "hard" criteria, relate to those criteria which are generally negotiable and do not counteract existing laws such as nature protection legislation.
12 The representatives of the citizens' initiative who participated in the focus group discussion wished to remain anonymous. They shall henceforth be named FS, AF and

TB and be referred to as "the members of the anti-wind group". The interviewed councilor is also a member of the group and will be referred to as "municipal representative" (or MR). The mayor of the small local municipality of Temnitzquell, Johannes Oblaski, who was interviewed together with MR, did not wish to be anonymous and is referred to as JO.

References

Amt Temnitz (2016a): Ergebnisprotokoll zu der 2. Sitzung des Amtsausschusses des Amtes Temnitz im Jahr 2016 am 2. März 2016. Available online at www.amt-temnitz. de/inhalte/amt_temnitz/_inhalt/politik-verwaltung/amtsauss/protokolle/proaa02.03.2016, checked on 11/27/2017.

Amt Temnitz (2016b): Temnitztal. Kurzporträt. Available online at www.amt-temnitz.de/ inhalte/amt_temnitz/_inhalt/startseite/startseite, checked on 9/9/2016.

Bafoil, François (Ed.) (2016): L'Énergie Éolienne en Europe. Conflits, Démocratie, Acceptabilité Sociale. Paris: Sciences Po les presses (Domaine Développement durable).

Forsa (2009): Umfrage zum Thema „Erneuerbare Energien" 2009 – Einzelauswertung Bundesländer. Available online at www.energieeffizient-sanieren.org/data/FORSA-Akzeptanz%20EE_Einauswertung%20Bundeslaender.pdf, checked on 5/20/2016.

Gegenwind Temnitz (2014): Aktionsbündnis Gegenwind Prignitz-Ostprignitz-Oberhavel gegründet. Available online at www.gegenwind-temnitz.de/?p=179, checked on 4/9/2015.

Gegenwind Temnitz (2016a): Über uns. Available online at www.gegenwind-temnitz. de/?page_id=52, checked on 9/9/2016.

Gegenwind Temnitz (2016b): Windkraft in Maßen, nicht in Massen. Available online at www.gegenwind-temnitz.de/, checked on 9/8/2016.

GL Berlin-Brandenburg (2016): Region Prignitz-Oberhavel. Gemeinsame Landesplanungsabteilung Berlin-Brandenburg (GL). Available online at http://gl.berlin-brandenburg.de/regionalplanung/regionen/strukturdaten-region-prignitz-oberhavel-398238.php, checked on 9/8/2016.

Jäger, Eva (2015): IRS Aktuell Meldungen: Über 80 Bürgerinitiativen gegen Windkraftanlagen in Brandenburg. Leibniz-Institut für Raumbezogene Sozialforschung. Available online at www.irs-net.de/aktuelles/meldungen-detail.php?id=304, checked on 10/13/2016.

Kuschel, Ansgar (2015): Beitrag der Regionalen Planungsgemeinschaft Prignitz-Oberhavel zu der SRL-Veranstaltung „Alle Lampen an!" am 8. Juli 2015 in Potsdam. Potsdam, 7/8/2015.

Landtag Brandenburg (2009): Volksgesetzgebung – Landtag Brandenburg. Landtag Brandenburg. Available online at www.landtag.brandenburg.de/de/mitgestalten/ volksgesetzgebung/396774, updated on 4/30/2009, checked on 9/28/2016.

Landtag Brandenburg (2014a): Antrag der CDU-Fraktion und der Abgeordneten Schülzke, Schulze und Vida. Verlässliche Abstandskriterien für Windkraftanlagen in Brandenburg. Drucksache 67/233.

Landtag Brandenburg (2014b): Antwort der Landesregierung auf die Kleine Anfrage 3409 des Abgeordneten Christoph Schulze, Fraktion Bündnis 90/Die Grünen. Entwicklung der Windkraft im Land Brandenburg. Drucksache 5/8731.

Landtag Brandenburg (2015a): Beschlussempfehlung und Bericht des Hauptausschusses zu der Volksinitiative nach Artikel 76 der Verfassung des Landes Brandenburg „Volksinitiative für größere Mindestabstände von Windrädern sowie keine Windräder im Wald". Drucksache 6/2593.

Landtag Brandenburg (2015b): Entschließungsantrag der CDU-Fraktion. Volksinitiative für größere Mindestabstände von Windrädern sowie keine Windräder im Wald. Drucksache 6/2643.

Landtag Brandenburg (2015c): Entschließungsantrag der CDU-Fraktion zum Gesetzentwurf „Gesetz zur Änderung der Brandenburgischen Bauordnung (Drucksache 6/1537, Neudruck). Drucksache 6/1671.

Landtag Brandenburg (2015d): Volksgesetzgebung in Brandenburg seit 1992. Übersicht über beim Landtag Brandenburg eingegangene Volksinitiativen gemäß Artikel 76 Landesverfassung sowie ähnlicher Anliegen. Landtag Brandenburg. Available online at www.landtag.brandenburg.de/de/mitgestalten/volksgesetzgebung/volksgesetzgebung_in_brandenburg_seit_1992/396602, checked on 9/28/2016.

Landtag Brandenburg (2016): Antrag der CDU-Fraktion. Energieland Brandenburg: Sicherheit, Berechenbarkeit, Akzeptanz. Drucksache 6/4219. Landtag Brandenburg. Potsdam.

Landtag Brandenburg (2019): Beschlossene Gesetze der 6. Wahlperiode – Landtag Brandenburg. Landtag Brandenburg. Available online at www.landtag.brandenburg.de/de/parlament/plenum/gesetzgebung/beschlossene_gesetze_der_6._wahlperiode/658313, checked on 1/10/2020.

MAZ (2016a): Umfrage zu Erneuerbaren Energien – Windräder in der Nachbarschaft? Kein Problem! In *Märkische Allgemeine* 7/3/2016. Available online at www.maz-online.de/Brandenburg/Windraeder-in-der-Nachbarschaft-Kein-Problem, checked on 7/11/2016.

MAZ (2016b): Volksbegehren Windkraftanlagen – Niederlage für Brandenburgs Windkraft-Gegner. In *Märkische Allgemeine* 7/6/2016. Available online at www.maz-online.de/Brandenburg/Niederlage-fuer-Brandenburgs-Windkraft-Gegner, checked on 7/11/2016.

MAZ (2016c): Kampf um hohe Netzentgelte – Woidke droht mit Windkraft-Stopp. In *Märkische Allgemeine*, 5/13/2016. Available online at www.maz-online.de/Brandenburg/Woidke-droht-mit-Windkraft-Stopp, checked on 5/13/2016.

MAZ (2016d): Weniger Windräder, weniger Stromkosten – Das bedeutet der Öko-Kompromiss für Brandenburg/Brandenburg – MAZ – Märkische Allgemeine. In *Märkische Allgemeine* 6/2/2016. Available online at www.maz-online.de/Brandenburg/Das-bedeutet-der-Oeko-Kompromiss-fuer-Brandenburg, checked on 6/3/2016.

Mehr Demokratie (2017): Mehr Demokratie e.V. Landesverband Berlin/Brandenburg: Übersicht Volksbegehren. Available online at https://bb.mehr-demokratie.de/bran-land-uebersicht.html, checked on 12/5/2017.

MW (2008): Energiestrategie 2020 des Landes Brandenburg. Ministerium für Wirtschaft des Landes Brandenburg. Potsdam. Available online at www.energie.brandenburg.de/media_fast/bb1.a.2755.de/Energiestrategie_2020.pdf, checked on 8/28/2012.

MWE (2012a): Energiestrategie 2030 des Landes Brandenburg. Potsdam.

MWE (2012b): Energiestrategie 2030 des Landes Brandenburg. Katalog der strategischen Maßnahmen. Potsdam.

MWE (2014): „Energiepolitischer Weg als richtig bestätigt". Minister Gerber zum Abschneiden Brandenburgs im Bundesländervergleich Erneuerbare Energien. Pressemitteilung 05.12.2014. Available online at www.mwe.brandenburg.de/sixcms/detail.php/bb1.c.383382.de, checked on 11/28/2014.

MWE (2016): Die Umsetzung der Energiewende aus Sicht der Landespolitik. Available online at www.prignitz-oberhavel.de/fileadmin/dateien/dokumente/REM/Regionalkonferenz/6_Praesentation_MWE.pdf, checked on 11/26/2016.

NHB (2012): Stellungnahme zum Entwurf der Energiestrategie 2030 des MWE vom Januar 2012. Available online at www.nachhaltigkeitsbeirat.brandenburg.de/cms/media.php/bb2.a.5490.de/en_str_stn.pdf, checked on 11/22/2016.

Overwien, Petra; Groenewald, Ulrike (2015): Viel Wind um den Wind. Aktuelle Heraus-forderungen für die Regionalplanung in Brandenburg. In Bundesinstitut für Bau-, Stadt- und Raumforschung (Ed.): Ausbaukontroverse Windenergie (Informationen zur Raumentwicklung, Heft 6.2015), pp. 603–618.

PIK; Raum und Energie; Universität Potsdam; Christian-Albrechts-Universität zu Kiel (2016): Energiekonflikte. Akzeptanzkriterien und Gerechtigkeitsvorstellungen in der Energiewende. Kernergebnisse und Handlungsempfehlungen eines interdisziplinären Forschungsprojekts.

RPG Havelland-Fläming (2014): Regionalplan Havelland-Fläming 2020. Tücken der Abwägung. Neuruppin, 6/17/2014.

RPG Prignitz-Oberhavel (Ed.) (2010): Regionales Energiekonzept für die Region Prignitz-Oberhavel. Kurzfassung August 2013. Available online at www.prignitz-oberhavel.de/fileadmin/dateien/dokumente/energiekonzept/REnKon_Endbericht_ Kurzfassung.pdf, checked on 11/22/2016.

RPG Prignitz-Oberhavel (2016): Planungsausschuss 3/2016. Neuruppin, 31. Mai 2016. Available online at www.prignitz-oberhavel.de/fileadmin/dateien/dokumente/ planungsausschuss/2016/03_2016/PA_03_2016_Praesentation.pdf, checked on 9/8/2016.

RPG Prignitz-Oberhavel (2017): Planungsgemeinschaft Prignitz Oberhavel: Regionalplan Freiraum und Windenergie. Available online at www.prignitz-oberhavel.de/planwerke/ regionalplan-freiraum-und-windenergie.html, checked on 11/8/2017.

SPD Brandenburg (2015): Akzeptanz der Windenergie sichern. Beschluss des SPD-Landesvorstandes vom 7. Juli 2015. Potsdam.

SPD Brandenburg; DIE LINKE Brandenburg (2014): Sicher, Selbstbewusst und Solidar-isch: Brandenburgs Aufbruch vollenden. Koalitionsvertrag zwischen SPD Brandenburg und DIE LINKE Brandenburg für die 6. Wahlperiode des Brandenburger Landtages 2014 bis 2019.

Tagesspiegel (2009): Brandenburger wehren sich gegen Windräder. In *2006*, 4/27/2009. Available online at www.tagesspiegel.de/berlin/brandenburgi/volksinitiative-brandenburger-wehren-sich-gegen-windraeder/1799454.html, checked on 9/28/2016.

5 Preserving health, curbing costs

The anti-wind movement in Canada, case study Ontario

Canada's most populous and economically most important province Ontario occupies the first place in Canada with regard to installed wind power capacity. Its 2009 Green Energy Act (GEA) introduced a feed-in tariff to support the development of renewable energy, but marked also a milestone for the evolving anti-wind movement. Both the discursive and institutional context prevalent in Ontario at the time, its discursive energy space, contributed to the movement's success in building alliances, shaping discourses and ultimately influencing policy. This chapter tells the story about the anti-wind umbrella organization Wind Concerns Ontario (WCO), their perspective on wind turbine development, the response the movement received from the main opposition party and how their demands were addressed by the government. A case study of a local anti-wind conflict is also presented, including the storylines of anti-wind grassroots organizations and their activities against a proposed wind project. This chapter is based on 32 interviews with representatives from civil society, government, companies and politicians at the township and provincial level as well as three focus group discussions with local anti-wind grassroots organizations that were conducted in Ontario in 2015. The chapter focuses on the time period from around 2010 to 2016.

After the GEA was implemented, wind energy opposition started to rise considerably and made wind turbines a debate unfolding at the provincial policy level. A July 2010 opinion poll found that 89 percent of Ontarians supported wind energy in their region of the province, although there were differences between the North where 93 percent approved, the Greater Toronto Area (91 percent) and the South West (84 percent) (Ipsos Reid 2010). "Loud/noisy/noise pollution" figured highest among the mentioned drawbacks of wind energy, just before "ugly/not pleasing to the eye/eye sore" and "lack of wind/no wind days/ requires wind" (Ipsos Reid 2010). Six years later, a poll suggested an even split in public opinion on wind turbines: 43 percent viewed Ontario's approach to wind energy in a positive and 43 percent in a negative way (Miner 2016). Nonetheless, a poll conducted in May 2016 found that 74 percent of Ontarians approved of the coal phase-out and the support of renewable energy as the right strategy for Ontario, and 81 percent wanted to see more renewable energy in the province (Environmental Defence 2016). These three polls show that support for wind energy has declined from 2010 to 2016, although the majority still approves of Ontario's general decision to phase-out coal and support renewables.

Anti-wind protest and public debate in Ontario

The predominant issues in the debate over wind turbines in Ontario are the concerns regarding human health, community benefits and economic impacts, including rising electricity costs (Baxter et al. 2013; Hill and Knott 2010; Songsore and Buzzelli 2014; Walker et al. 2014a, 2014b).[1] The fact that planning power was taken away from the municipalities is regarded as a major source of protest (Bues 2018; Walker and Baxter 2017). Despite this streamlined procedure implemented with the GEA, each wind developer must carry out consultation processes with the local population in order to acquire a Renewable Energy Approval. The following section recapitulates the viewpoint of the CEO of an Ontario-wide consultancy firm (BH) and one of his employees (JZ) (Anonymous interview consultants BH and JZ). Having carried out public consultancy processes for a multitude of wind energy developers in southern Ontario, JZ and BH call the anti-wind movement in Ontario "hugely fervent", "energetic" and a "crusade" (Interview, consultants BH and JZ). JZ reports that in some parts of the province, especially in Central Ontario, she has experienced each public consultation meeting as very contentious and recalls that one of her projects was cancelled twice because the landowners backed out of the project after they had been threatened and had their property vandalized. Having carried out public consultation processes regarding other land use and environmental planning issues, BH describes how he has perceived the anti-wind movement in Ontario:

> What has amazed me is this hard-biting, almost religious fervor of that movement, and you don't see that every day. So we have other debates in the public on other issues. We have debates on education, public transit, healthcare, people have views on that. People agree and disagree on those subjects. But we do not have 800 people with picket signs and tractors, we do not have vandalizing their neighbor's property (Interruption JZ: they don't threaten their neighbor's kids) because they have taken one position over another. It is almost like the, I am thinking of Ireland, the Catholics and the Protestants fighting each other over deeply held religious beliefs and being prepared to become aggressive, be prepared to take non-lawful measures, and we have had all of that in Ontario, which is an extremism in how far it has gone that it is- that is what makes it so bizarre.
>
> (Interview, consultant BH)

This quote exemplifies the intensity of anti-wind protest from a consultant's perspective. The dispute over wind turbines in the vicinity of a proposed project is described as a conflict pitting neighbors against neighbors – between those who are in favor of turbines and those who are not. At the consultation meetings of proposed developments, the consultants experienced forceful rallies of numerous opposing residents, leading BH to compare the force of anti-wind disputes in Ontario to violent religious conflicts. Both consultants further explain the anti-wind dispute as largely resulting from an information deficit and the fact that people do not trust any material provided by the developers. JZ reports that the scientific evidence they put forward during the open house meetings to address the residents' concerns was not trusted and "basically just thrown right out the window" (Interview, consultant JZ). In the opinion of both consultants, the industry should have engaged right from the beginning and on a greater scale in education, information, dialogue and communication. Also, the government should have been more active in this regard, because the information from the industry was often perceived as biased. Consultant BH believes that this vacuum of trusted information left by government and industry left it open to the anti-wind movement to "instill those fears" that helped to "fuel the movement" (Interview, consultant BH). Indeed, anti-wind sentiments in Ontario revolve to a large degree around the question of which information is to be trusted, particularly with regard to health concerns.

Health concerns[2] have had a prominent place in anti-wind storylines in Ontario since the first commercial wind turbines were set up in Ontario in 2002, but they have amplified in quantity and quality during the GEA implementation period. This has also been transported by the media. The number of Ontario-focused news stories regarding adverse health and noise impacts of wind turbines has skyrocketed during the lead-up to the GEA in late 2008 and early 2009 (Hill and Knott 2010). Prior to the GEA, health concerns as reported in the media mainly revolved around uncertainties regarding the health effects of turbines and protestors expressed a general trust in science (Songsore and Buzzelli 2014). After the enactment of the GEA, those concerns were more affirmatively put forward by dismissing existing scientific evidence that claimed turbines to be safe. This was legitimized by testimonies from people living close to turbines and by referring to other jurisdictions that engaged in detailed health studies (Songsore and Buzzelli 2014). Furthermore, media articles before the GEA revealed an extreme distrust in the Ontario government to protect citizens from potential health hazards, and while letter writing had been a major form of protest activity before the enactment of the GEA, the period after the GEA saw a more "radical engagement" in the form of rallies and petitions (Songsore and Buzzelli 2014). This corresponds to the testimonies of the two consultants presented earlier regarding the confrontational consultation meetings they had experienced.

The media not only conveyed this shift in peoples' attitudes on health, but were also found to associate wind turbines with fear-inducing characteristics. An analysis of the media coverage of five Ontario wind energy facilities between 2007 to 2011 shows that 94 percent of the analyzed newspaper articles contained the fright factor "dread" which the authors of the study define as "negative, loaded or fear-evoking description of health-related signs, symptoms or adverse effects of wind turbine exposure" (Deignan et al. 2013, p. 239). Furthermore, 58 percent of the newspaper articles associated wind turbines with the descriptions of "poorly understood by science", 45 percent with "involuntary exposure" and 42 percent with "inequitable distribution" (Deignan et al. 2013, p. 239). After the GEA was implemented, the total number of occurrences of each fright factor increased, out of which "dread" and "poorly understood by science" increased significantly (Deignan et al. 2013, p. 239). The authors conclude that this representation may "produce fear, concern and anxiety" (Deignan et al. 2013, p. 234) for readers of media articles. Next to this representation of wind turbine development in newspapers, there were also documentaries and talk shows on public television that raised the issue of whether there were adverse health concerns to be expected from wind turbines (CBC 2015; TVO 2014, 2015).

Health concerns regarding wind turbines were not only raised in the Province of Ontario. In response to growing concerns, the federal ministry for national public health, Health Canada, announced a large-scale epidemiology study on wind turbine noise and health in 2012. A summary of the results was released in late 2014. Self-reported illnesses, sleep disorders and stress were not found to be associated with turbine exposure. Nonetheless, annoyance with wind turbine features including noise, blinking lights or vibrations were found to be statistically associated with increasing levels of wind turbine noise (Government of Canada 2014). As there were two years between the announcement of the study and the release of a summary of its findings, a common demand of the anti-wind movement was a temporary moratorium on the further construction of wind turbines until publication of the results of the Health Canada study. The summary of the results was contested by the anti-wind umbrella WCO, criticizing the methodology used and arguing that they provided contradictory results (Wind Concerns Ontario 2014). The next section has a closer look at how WCO emerged and became an important actor in disputes over wind turbines in Ontario.

Scaling-up the debate: wind concerns Ontario

The umbrella organization WCO played a key role in linking both the emerging grassroots groups with each other as well as upscaling their concerns to the provincial level. As of November 2015, WCO counted 49 member organizations from all over rural Ontario. The organization's demands with regard to wind turbines are described on their website as follows:

- No new FIT contracts or project approvals
- Cancel projects with FIT contracts not yet approved
- Enforce existing audible noise standards
- Update regulations to respond to Health Canada data and develop regulations related to low frequency noise and infrasound immediately
- Compensate for the impact of turbine operations on neighbours

(Wind Concerns Ontario 2015)

This list shows that their demands either relate to the cancellation of projects or to health-related issues, which reflects the dominance of the health frame in disputes over wind turbines in Ontario. In order to characterize the anti-wind movement in more detail, the following section presents the development of the umbrella organization WCO with a particular emphasis on its first president, who built the group into an anti-wind umbrella organization active throughout Ontario (Interview, John Laforet).

WCO started in October 2008 as a loose information-sharing alliance focusing on on-shore wind energy. When the then 23-year-old student John Laforet became their president in August 2009, WCO became a strategically focused interest organization which benefitted from the political experience of its new president (Interview, John Laforet). John Laforet had first become involved in a local anti-wind group in 2008 when off-shore wind turbines were proposed off the Scarborough Bluffs in the Scarborough district of Toronto. Initially, as John Laforet states, he was in favor of the city producing its own electricity because the city of Toronto relied on the rural areas for many other issues, such as growing food, storing garbage or generating electricity. His motivation to join the local anti-wind group was concern about the way in which information was provided, but he was not convinced of the "wind turbine syndrome"[3] and did not believe in a real threat to the environment. Instead, he initially "got involved because it was not fair" (Interview, John Laforet). As John Laforet describes, the early protest in Scarborough started as a local risk issue. One of the major concerns related to the unknown effects of vibration from the wind turbines travelling through the rock towards residential houses. Those who raised this topic did not feel taken seriously, not being provided enough information and felt themselves forthrightly dismissed as "NIMBY". The fact that the then premier of Ontario Dalton McGuinty introduced the GEA in 2009 as "stopping NIMBYs" while specifically waving at the Scarborough group drove many people even further into an attitude of resistance, according to John Laforet. Furthermore, as John Laforet states, Scarborough has been known as an area of a strong identity with their escarpment and its residents had fought other developments before. Regarding the proposed wind turbine development, John Laforet reports:

So the very initial concern was what will happen. It was the unknown, the fear was the unknown. (…) so in 2008, the perspective on Wind Turbine Syndrome was, if this was real, it is very concerning because what happens to you if your house is worthless and you get sick by living there? So that was a big concern. So that was the issue and being called NIMBY, being attacked for not having the answers and not having been given the answers, and so within the region, not Ontario, but within the general area, Love Canal is a very famous case of people being poisoned by the land and they were called NIMBYs and they are just across the lake. People older than me remember when that happened, I have heard about it, but the sixty year olds in the group were alive when that was happening, and so the view of NIMBY to someone who is aware of that is very different, because it is – you know. So those guys are like, birth defects, weird cancers, all kinds of stuff, because they are living on toxic waste.

(Interview, John Laforet)

The comparison to Love Canal[4] underscores the deeply felt view in the group that building wind turbines in the lake represented a severe health threat. It compares to living next to a toxic waste dump and involves the possibility of becoming a new instance of Love Canal. The reference to Love Canal "reached into it as far as proof that when somebody calls you NIMBY, they might be hiding something" (Interview, John Laforet). Love Canal is thus regarded as a reminder of not giving up in the face of being dismissed as following one's own interest only.

Under John Laforet, WCO benefitted from his former political experience as acting executive assistant to Brad Duguid (who later became Minister of Energy) and elected riding director of the Ontario Young Liberals. Drawing on his previous experience in community organization, under John Laforet WCO followed a three-fold strategy, including the local level, the provincial level and the media. First, WCO aimed at the local level by encouraging rural Ontarians to get actively involved in the anti-wind struggle by joining rallies, delivering testimonies to the local press and asking their councils and Members of Provincial Parliaments (MPPs) to pass or support motions for a moratorium on wind turbines until health studies were completed (Interview, John Laforet). Furthermore, local groups were encouraged and supported to appeal each proposed wind turbine project before the Environmental Review Tribunal. Second, their strategy was also geared towards provincial politics. Aiming at eliminating the GEA as a whole, their 'natural' ally was soon to become the Progressive Conservative Party (PC, discussed later). The two strategic topics of the campaign were "democracy", referring to the superseding of the local level in the approvals process, and "health", referring to audible and inaudible noise:

Because if the debate was coal or wind, we are not winning that fight, not in a million years, and that is not a fight you should win, but it is also a false argument. If you go down that rabbit hole, you will lose. If we talk about our rights, we should win and be able to garner some sympathy, which will get people to listen to us when we say, "And there are these other problems with renewables". So, fundamentally we framed it that way.

(Interview, John Laforet)

As this quote exemplifies, the campaign was strategically focused on "rights", which was believed to receive broad support. The framing of their struggle was about "science and democracy because those are pretty safe" (Interview, John Laforet), as those represent robust issues in society that hardly anyone was believed to be opposed to. This deliberate decision underscores the strategic aspect behind the anti-wind campaign with the aim to unite diverse actors and interests behind the common goal of fighting the GEA. John Laforet's previous experience in politics also helped him build a media strategy for WCO:

So one thing that I brought to the fight was a stronger emphasis on the need for media to educate, and because we did not have- like no money for advertising or anything, so you need journalists to tell your story for you, and they are more believed anyway, so it works really well. And so Wind Concerns Ontario took a very prominent role in telling the story of what was happening in Ontario and we asked local groups to tell their stories to local media and work with municipal leaders to tell local media what was wrong and get MPPs to tell local media. So that was a big part of it, it was building a robust media strategy that involved a central message with 58 variations because it meant you could be in the news somewhere in Ontario absolutely every day.

(Interview, John Laforet)

This comprehensive media strategy involved joining forces with different actors including anti-wind groups, municipalities, First Nation groups and MPPs to contribute to their media strategy based on their common emphasis on "rights". John Laforet was soon known in Ontario as the leader of the anti-wind movement and reports to have been called by hedge funds from the US who wanted to know whether they could invest in a particular project and as a matter of fact, he encouraged them to refrain from it. John Laforet summarizes the overall strategy of the campaign:

So the goal was dry up the money, scare away the investors, delay the proponents, and attack the government, and if all of that works, nothing will get approved, and then we get to the election and the government loses and then Tim Hudak comes in and 98 percent of the projects will never be built. So that was the plan. Three of the four pieces worked.

(Interview, John Laforet)

As this quote exemplifies, the campaign found its peak in the run-up to the 2011 elections in which the leader of the opposition party Tim Hudak played an important role in making localized anti-wind struggles a provincial discussion. In the months preceding the elections, John Laforet toured the rural areas on a "Truth About Turbines Tour" to organize and support anti-wind protest. In John Laforet's view, WCO's and their local groups' protest led to the postponement of many wind projects, but the fourth goal of overturning the Liberals in the next election did not become a reality.

The opposition party as ally

Due to their importance in terms of electoral support and the fact that they were in government in Ontario for a number of legislative terms, the Progressive Conservatives (PC) represented an important potential candidate for supporting the anti-wind movement (Walker et al. 2018). The PC advocates a free market economy and rejects subsidies for any form of electricity generation, including renewables. Hence, the PC frames electricity mainly in terms of "cost" (PC Caucus 2012). This section explores the support the PC has granted to the anti-wind movement from the viewpoint of its then leader Tim Hudak (Interview, Tim Hudak). Tim Hudak, who became party leader in June 2009, recalls that the PC caucus was initially divided as to what degree they should oppose the GEA, because "certainly the principle of more environmentally friendly power was a positive one" (Interview, Tim Hudak). Informed by a study the PC had commissioned that demonstrated an increase in electricity rates as a consequence of an introduction of the FIT, the PC eventually opposed the GEA, although at that time, "the vast amount of public opinion was supportive of the Act and we felt a bit like voices in the wilderness" (Interview, Tim Hudak). The PC based their criticism of the GEA mainly on the cost issue and a bit less on the overriding of local decision-making.

As a newly elected party leader, Tim Hudak travelled the province and soon became aware of the growing public disaffection regarding wind turbines. He reports that wind turbines were identified as a major issue in many communities, which became evident both by a previous scan of the local media and during meetings with mayors, business leaders or citizens directly at townhall meetings:

It was also common for me to do townhall meetings where anybody in the community could come and listen to me and ask me questions, and I saw more and more in rural Ontario larger and larger groups of people opposed to the wind turbines, and strongly opposed. You get an ability to read people, are they passionate, are they emotional, are they very upset or was it an academic or a heartfelt feeling? And you could just see the temperature rising on the wind issues in many dispersed pockets across our province and then the hydro rates[5] were rising, you saw a general concern about the price of energy in our province and its impact particularly on people of low incomes and on the competitiveness of manufacturing.

(Interview, Tim Hudak)

This quote exemplifies the way in which the PC leader experienced the rising strength of anti-wind sentiments in Ontario, both in number of people opposing wind turbine development as well as in the rigor of putting forward their arguments. Tim Hudak also connects this rising contestation with the increase in electricity rates in Ontario during that time.[6] According to Tim Hudak, the cost issue was among the main topics of concern, along with the feeling of being overruled and the fact that some neighbors benefitted while the rest "have to suffer the consequences" (Interview, Tim Hudak). The reported concerns, as Tim Hudak states, included impacts on the local environment and on peoples' health, on the enjoyment of their property and property values. He agrees with the criticism of superseding the municipal level and regards the ERT as having "been changed from honest courts based on science and good argument into rubberstamping bodies for the wind turbines" (Interview, Tim Hudak).

As the leader of the PC fully endorsed the arguments raised by the increasing number of citizens opposing wind turbine development, he supported them in a number of ways. The PC tabled several motions to call for a moratorium on wind turbines until health studies were undertaken. The call for a moratorium became an important issue that Tim Hudak placed on his platform for the 2011 provincial elections (see also Cross et al. 2015). Furthermore, he met with the leader of WCO John Laforet on several occasions to discuss communication politics:

I had known him before he took this job and then meeting in my office on several occasions to talk about what we could do to stop this, how we can highlight the issue, how we could connect the issues happening in rural Ontario with people in urban areas so they would understand the negative impacts, so we had a lot of discussions about the communications policy.

(Interview, Tim Hudak)

This strong cooperation between PC and the anti-wind umbrella organization was most intense in the run-up to the 2011 elections. While the president of WCO toured the rural areas to convince local residents to vote for the PC via the "Truth About Turbines Tour", specifically targeting Liberal ridings that had turbines (Lees 2011), the PC opposition leader ran on a platform for a moratorium on wind turbines. The PC leader reports that people were asked to vote for PC not for their conservatism, but because they were the only party supporting the anti-wind movement:

And then we were the voice for the anti-wind side. We were the only party that took that position. I did a series of press conferences, questions in the legislature, the moratorium bill. Tried to galvanize the anti-wind sentiment across the province. And those ridings that had major wind projects, almost all of them, probably not all but almost all of them voted conservative in the next election campaign.

(Interview, Tim Hudak)

As exemplified by this quote, Tim Hudak engaged in a variety of supportive actions towards uniting the anti-wind movement and even described the PC as the "voice" for the anti-wind movement. The quote also hints at the outcomes of the October 2011 provincial elections. The Liberals lost their majority by one seat and won a minority government. The media attributed the Liberal's loss of 18 seats mainly to the issue of wind turbines (e.g. Warren 2013). A study showed that the incumbent provincial government's vote share indeed declined by five percent in ridings with wind turbine proposals and by ten percent in ridings with operational wind turbines (Stokes 2016).

The dispute over wind turbines thus had an effect at the ballots. Tim Hudak is convinced that "as a result of our actions, there are fewer wind turbines in the province" (Interview, Tim Hudak). He further states that even if they were not able to place the moratorium, they were able to raise the issue "successfully in the media" (Interview, Tim Hudak). The fact that disputes over wind turbines had gained unprecedented levels by that time is also exemplified by the changing position of the Ontario Federation of Agriculture (OFA). The OFA represents a traditionally important interest group in Ontario politics and had been a member of the GEA Alliance. In early 2012, the OFA issued a statement that it would no longer support wind turbines:

> However, the situation regarding Industrial Wind Turbines (IWT) has become untenable. The proliferation of wind turbines across rural Ontario has seriously polarized our rural communities. Residents not engaged in turbine developments have been pitted against neighbours, over concerns with health impacts and quality of life issues. IWT development currently preoccupies the rural agenda.
>
> (OFA 2012)

It becomes evident that by calling wind turbines "industrial", the OFA adopted the language of the anti-wind movement and put forward their raised concerns such as adverse health impacts and quality of life. The statement also highlights that wind turbines "preoccupy" the rural agenda. To sum up, the anti-wind movement in Ontario worked closely together with the Progressive Conservatives as the main opposition party with the result that even the OFA as an initial supporter and member of the GEA Alliance changed its viewpoint. The issue of wind turbines had now become fully present at the provincial policy level.

Government response to protest and GEA changes 2012–2016

The concerns that the anti-wind movement together with the Progressive Conservatives (PC) raised were the superseding of the local level, the cost issue, and concerns for human health. These main topics were expressed in WCO's emphasis on "rights" during their campaign, the PCs focus on "costs" and both WCOs and PCs call for a moratorium of wind turbines until health studies were completed. This section turns to how the government reacted to these demands. It must be noted that government response was not exclusively directed at anti-wind protestors. Especially the government's reaction towards reducing the prices paid by the FIT program addressed a debate on costs that had unfolded in the general public and at the provincial politics level (see also Stokes 2013).

The Ontario government's first reaction to anti-wind protests occurred with the implementation of the GEA's decision-making system, which superseded local input. The then Premier Dalton McGuinty introduced the act by referring to "overcoming Nimbyism", although addressing protest had only been one motivation for it. After this very initial response, two main types of government responses to criticism against the GEA can be identified: a response focused on information regarding health impacts and one focused on substantive policy changes in 2013 and 2016.

Addressing health concerns

Adverse health impacts have been one of the core topics of disputes over wind turbines in Ontario. Even before the GEA was implemented, this discussion had reached the provincial level. As a reaction, the Government of Ontario, in the spring of 2009, announced five years of funding for the "Ontario Research Chair in Renewable Energy Technologies and Health" to study the health and safety implications of renewable energy generation (Council of Ontario Universities 2010; Hill and Knott 2010). In an October 2009 memorandum, Ontario's Chief Medical Officer of Health Dr. Arlene King stated that "in response to public concerns", ministry staff had reviewed the literature on the potential health impacts of wind turbines, but had found that the literature could not establish a link between wind turbine noise and adverse health effects (King 2009). The report released in May 2010 concluded that

> while some people living near wind turbines report symptoms such as dizziness, headaches, and sleep disturbance, the scientific evidence available to date does not demonstrate a direct causal link between wind turbine noise and adverse health effects. The sound level from wind turbines at common residential setbacks is not sufficient to cause hearing impairment or other direct health effects, although some people may find it annoying.
>
> (CMOH 2010, p. 3)

As this quote highlights, the annoyance factor of wind turbines is acknowledged, but self-reported adverse health symptoms were not confirmed to be associated with wind turbine noise. The government of Ontario therefore addressed the concern regarding human health by conducting a literature review and providing research funds. Thus, the government did not place a moratorium and did not change their policy or implementation procedure of wind turbines in response to health concerns. By contrast, the "rights" and "cost" concerns were addressed by major policy changes.

Addressing "rights" and "costs": substantial GEA policy changes

Soon after the GEA was passed into law, the FIT prices paid for solar were reduced, responding to criticism regarding the relatively high initial prices paid. In February 2011, the government put a moratorium on off-shore wind energy and cited environmental concerns. The ministry stated that offshore wind turbines in freshwater environments were early in development and needed further scientific studies (Taylor 2011). Existing contracts for planned developments were cancelled, which one affected company brought before the NAFTA court. It ruled that the government had to pay CAD 25 million in damages for inequitable treatment to the plaintiff company, finding that Ontario's decision was "at least in part" motivated by a genuine concern about a lack of scientific data but was also affected by public opposition and the Liberal's corresponding concerns regarding the October 2011 elections (Jones 2017).

It was under the impression of the 2011 election results that the previously scheduled review of the FIT began. The review stated that more than 2,500 small and large FIT projects had been approved and that the FIT had attracted more than CAD 27 billion in private sector investment, welcomed more than 30 clean energy companies, created more than 20,000 jobs and was on track to creating 50,000 jobs (Ministry of Energy 2012b, p. 2). The review was carried out in the form of a large consultation process including submissions by 2,900 individuals and organizations and the report was published in March 2012 (Ministry of Energy 2012b). In a news release, the Ontario Ministry of Energy announced they would act on the review's following key recommendations:

- Creating more jobs sooner by streamlining the regulatory approvals process for projects while maintaining the highest environmental protection standards.
- Reducing prices – for solar projects by more than 20 percent and wind projects by approximately 15 percent.
- Encouraging greater community and Aboriginal participation through a new priority point system, which will also prioritize projects with municipal support.
- Reserving 10 percent of remaining capacity for projects with significant participation from local or Aboriginal communities.
- Developing a Clean Energy Economic Development Strategy to leverage Ontario's significant expertise and strengths to become a global leader in the sector.

(Ministry of Energy 2012a)

The review therefore addressed the "costs" concern by reducing prices and also the "rights" concern by improving local participation.[7] For this purpose, a new priority points system and a capacity carve-out reserved for projects with local participation was anticipated. Acknowledging that community participation "is important to the continued success of the FIT Program" (Ministry of Energy 2012b, p. 15), the priority points system ranks application for FIT contracts according to their number of gathered points. Community or Aboriginal participation adds priority points to the application, which increases the chances of receiving a FIT contract (Ministry of Energy 7/11/2012). The 2012 review thus resulted in changes to the FIT that intended to reduce costs and better incorporate the municipal voice. At the same time, the review held on to the basic characteristics of the FIT.

More substantial changes were undertaken in 2013. At that time, many of the original key actors behind the GEA had stepped down from their political function in provincial politics. After Premier Dalton McGuinty resigned for a controversy around cancelled gas plants,[8] Kathleen Wynne became the new Premier of Ontario in early 2013. With Dalton McGuinty as one of the initiators of the

GEA, Ontario politics lost one of its major renewable energy advocates. His Energy Minister Chris Bentley resigned for the same causes. Bentley's predecessor George Smitherman, Ontario's Minister for Energy and Infrastructure under whom the GEA had been launched, had already resigned in late 2009 to enter municipal politics. Shortly before Kathleen Wynne officially became the Premier of Ontario, she was cited in a newspaper article that she did not believe in veto power for municipalities regarding wind turbine siting, but proposed instead, "let's do it where the community is going to benefit, where there's been a good community process and we've got a willing host" (Martin 2013). This reference to a "willing host" sparked a province-wide reaction from the anti-wind movement who called upon municipalities to pass motions to declare themselves as an "unwilling host". As of December 2015, more than 90 townships and municipalities in Ontario have passed such a motion (Unwilling Hosts 2015). The following changes in the GEA thus have to be regarded in the light of the new actor constellation following McGuinty and his energy ministers' resignation as well as the rise of the "unwilling host" movement, which increased the visibility of municipal discontent over wind turbines in media articles and beyond.

In late May 2013, the Ministry of Energy announced the discontinuation of the FIT for renewable energy projects over 500 kW and its replacement by a new competitive procurement process. The threshold of 500 kW practically terminated the FIT for commercial wind power projects. Although competitive procurement schemes are usually associated with better cost control, the ministry's news release explained the decision by improving local control in renewable energy development. Headlined "Ontario working with communities to secure clean energy future", the announcement promised a new competitive procurement process that would "better meet[s] the needs of communities" by directly working with municipalities to identify appropriate locations (Ministry of Energy 2013b). The following three further changes were anticipated:

- Revise the Small FIT program rules for projects between 10 and 500 kW to give priority to projects partnered or led by municipalities.
- Work with municipalities to determine a property tax rate increase for wind turbine towers.
- Provide funding to help small and medium-sized municipalities develop Municipal Energy Plans – which will focus on increasing conservation and helping to identify the best energy infrastructure options for a community.

(Ministry of Energy 2013b)

These measures show that next to the existing priority points system, a further strengthening of municipal participation was envisioned by investigating the possibilities for better financial participation and the generation of municipal energy plans. These aim at developing a comprehensive municipal strategy to align infrastructure, energy use, and land use planning in order to "ensure that Ontario builds energy infrastructure in a process that respects communities"

(Ministry of Energy 2013a). The Ministry of Energy advised the Ontario Power Authority (OPA) in a corresponding directive to develop a competitive process for projects larger than 500 kW and "engage with municipalities to help inform the identification of appropriate locations and siting requirements for future renewable energy projects" (Ministry of Energy 2013c, p. 3). While 900 MW of new capacity was announced for small renewable projects until 2018, the precise procurement targets for the new competitive process were determined in the revised Long Term Energy Plan.

In December 2013, the new Ontario Long Term Energy Plan (LTEP) was released. The plan was the result of a consultation process comprising the view of almost 8,000 people on the topics of conservation, energy supply, regional planning and energy imports in an online survey (Ontario Ministry of Energy 2013, p. 58). Titled "Achieving Balance", the plan did not set new targets for wind, solar and bioenergy but extended the year for reaching the previous target of 10,700 MW from 2018 to 2021. The existing targets for hydro-electric were increased from 9,000 MW to 9,300 MW by 2025. By 2025, renewable energy including hydro-electric would represent about half of Ontario's installed capacity, amounting to 20,000 MW (Ontario Ministry of Energy 2013, p. 6). Regarding the new competitive procurement scheme for renewable projects larger than 500 kW, the LTEP planned to make available up to 300 MW of wind, 140 MW of solar, 50 MW of bioenergy and 50 MW of hydroelectric capacity in 2014 (Ontario Ministry of Energy 2013, p. 33). Two more procurement rounds were anticipated for 2015 and 2016 for which procurement targets would be determined later. The other half of Ontario's electricity generation would continue to be supplied by nuclear power. While the construction of new nuclear generating units was deferred because Ontario's energy demand had proven to be lower than expected, the plan commits to the nuclear refurbishment of the existing Darlington and Bruce Generating Stations. This had "received strong, province-wide support during the 2013 LTEP consultation process" (Ontario Ministry of Energy 2013, p. 29).

While the 2013 LTEP mentions a lower forecast in electricity demand for Ontario, the plan does not directly reveal why the targets for non-hydro renewables are not extended but postponed. This decision can however be explained by the importance the LTEP assigns to wind power. As "a clean, reliable energy system relies on a balance of resources" (Ontario Ministry of Energy 2013, p. 38), wind power is considered one possibility to diversify the province's electricity mix. Wind power is regarded as an important contribution to replace coal-fired electricity generation, which adds to "smog, pollution and climate change" (Ontario Ministry of Energy 2013, p. 38). The LTEP thus generally endorses wind power but regards it as a diversification and additional source of power rather than a central technology to base Ontario's electricity supply upon. In this vein, the LTEP endorses the new competitive procurement process as a way of awarding contracts to "cost-efficient and well-supported projects" (Ontario Ministry of Energy 2013, p. 33). While the 2013 official announcement of the FIT's replacement by a competitive procurement scheme for larger projects only referred to the benefit of increasing municipal participation, this statement in the LTEP also considers the costs perspective of the new process. The reasons why the LTEP does not extend the target for non-hydro renewables are thus indirectly justified by a lower forecast in electricity demand and the commitment to refurbishing existing nuclear generating stations against the backdrop of no general intention to make non-hydro renewables a central pillar of Ontario's electricity mix.

The new competitive procurement scheme (LRP: Large Renewable Procurement) was eventually organized in two rounds. The first began in 2014 and concluded in 2016. It procured 454 MW of new wind, solar and waterpower capacity in 16 contracts (IESO 2017b). The second process began in March 2016 but it was cancelled before procuring new capacity. In September 2016, Energy Minister Glenn Thibeault announced the suspension of the LRP program altogether, cancelling plans to sign contracts for up to 1,000 MW of power from renewable sources[9] (CBC News 2016a; Ministry of Energy 2016b). He argued that this would save the province up to CAD 3.8 billion in costs from its 2013 LTEP and that Ontario already had a strong supply of clean power. The emphasis on costs had therefore gained force at the provincial policy-making level. The revised 2017 LTEP, titled "Delivering Fairness and Choice", primarily focuses on measures to reduce electricity bills. Those build on already implemented measures including suspending the LRP's second round, postponing the refurbishment at the Bruce Nuclear Generating Station for four years, and deferring the construction of two new nuclear reactors at Darlington (Ontario Ministry of Energy 2017). In order to reduce electricity supply costs and increase flexibility, Ontario would also move away from long-term electricity contracts and instead enhance its marked-based approach. The plan generally regards Ontario's energy sector as in a good shape to meet electricity needs for the next 20 years. Apart from the previously decided refurbishment of nuclear generating stations, no other procurement targets are announced.

The 2017 LTEP marks the end of the government's ambitions to increase the rate of renewables. While fully endorsing the renewable sector in Ontario and its continuing economic benefits to the province, its status quo is regarded as sufficient for the next 20 years. When the existing renewable energy contracts expire, renewable energies are considered to be cost-effective and therefore considered likely to continue their operation (Ontario Ministry of Energy 2017, p. 43). By effectively halting further wind energy procurement, the government has largely taken over the concerted demands of the anti-wind movement and the political opposition to decrease costs and re-establish local planning authority. While the call for a moratorium on further wind turbines until health studies are completed was not directly taken over, the 2016 decision to halt the LRP process effectively yielded to that demand. It must be noted however that the government's reaction officially followed the consultation processes of the FIT and LTEP review and was not depicted as a direct reaction of the anti-wind movement's demands. In hindsight, Ontario's renewable energy program under the GEA was highly successful in establishing renewables other than hydro-electric as a firm constituent of Ontario's energy mix. In the long run however, wind, solar and biogas could not assert themselves against the strong role of Ontario's nuclear power sector and the rising emphasis on costs in the debate over energy.

The local level: a case study of Niagara region

This section turns to the local level and introduces the storylines of a local anti-wind grassroots organization, two township councilors, two township planners and one regional planner. The focus is on two municipalities in southern Ontario: West Lincoln and Wainfleet in Niagara Region. Both municipalities had wind turbines before a major project of 77 wind turbines was proposed to the area. Shedding light on this local level case study helps to understand and to illustrate the origins and characteristics of local anti-wind grassroots organizations and their discursive context in Ontario.

The Regional Municipality of Niagara (short: Niagara Region) is situated in southern Ontario, bordering the municipality of Hamilton to the northwest, Haldimand County to the southwest, Lake Ontario to the north and Lake Erie to the south, and the USA to the east. Home to the famous Niagara Falls, which produce hydro-electric renewable energy next to being a popular tourist destination, the region has declared itself as "green energy capital of Canada" (Bolichowski 2012). Niagara Region has a distinct agrarian character with few urban centers and comprises twelve local municipalities. In July 2011, the Niagara Region Wind Corporation (NRWC) publicly announced its plan to establish a 230 MW wind turbine project in the area of Niagara Region and adjacent Haldimand County (NRWC 2011). The private developer held several public meetings as legally required and received its official license for construction and operation, the Renewable Energy Approval, in 2014. Since, the project has been sold several times and consequently changed names. The project consists of 77 wind turbines with hub heights of 124 and 135 meters (Government of Ontario 2014) and was inaugurated in June 2017. While the proposed project locations for the wind turbines also cover the area of Haldimand County, the focus here is on the two municipalities of West Lincoln and Wainfleet of Niagara Region for which 44 turbines (West Lincoln) and two turbines (Wainfleet) were proposed. This section focuses on the time between 2011 and 2015, before the project was built. Table 5.1 provides basic data on both municipalities from the latest 2011 census.

Table 5.1 West Lincoln and Wainfleet: basic data

	Surface area in square kilometers	*Population*	*Population density (individuals per square kilometer)*
West Lincoln	388	13,800	35.7
Wainfleet	217	6,300	29

Source: Statistics Canada 2011a, 2011b.

Both municipalities were hosts to five wind turbines each before the new project was proposed. The earlier projects were granted their contract in 2010, their construction started in 2013 and they generated electricity one year later (Anonymous interview project developer). There were anti-wind protests in both West Lincoln and Wainfleet against this first project. A developer who was involved in both projects reports that anti-wind protest was more vocal in West Lincoln than Wainfleet (Anonymous interview project developer).

Both Wainfleet and West Lincoln council display strong opposition against both the older and the new wind turbine projects. In 2011, Wainfleet Township Council voted unanimously to call upon the provincial government to enact a moratorium until third-party studies on the environmental, economical and health impacts were conducted (Niagara This Week 2011). West Lincoln passed a similar moratorium against the installation of wind turbines until the completion of health studies (WLWAG 2011). In 2013, both townships followed the example of other Ontario municipalities and adopted a council motion that declared their unwillingness to become hosts to wind turbines. The motion in West Lincoln cites as reasons to oppose wind turbines in its municipality "a lack of information on long-term health effects, potential negative impacts on property values and long-term negative economic implications to the community resulting from the two applications" (Moore 2013). The Region of Niagara supported West Lincoln's and Wainfleet's decision but no majority was found to declare the whole region as such (Bolichowski 2013).

When NRWC proposed a CAD 12 million share Community Vibrancy Fund to West Lincoln, the council rejected the offer.[10] The fund is a common way in Ontario to have the municipality financially participate in the development of turbines. In a newspaper article, Mayor Doug Joyner was cited as saying "We're trying to be consistent. We're trying to say to the NRWC and to the province that we're not willing hosts" (Dakin 2013). Wainfleet council attempted to restrict turbine planning in their jurisdiction by enacting a bylaw that prescribed a standard turbine setback of 2 km. The regulation was brought before court where it was struck down. A Wainfleet councilor reports that the municipality was the first community in Ontario to test the adoption of that particular bylaw (Interview, CWF).

The way in which wind turbine planning is carried out in West Lincoln and Wainfleet reflects the typical procedure of wind turbine planning in Ontario. The main lines of communication take place between the private developer and the provincial government. After NRWC acquired the feed-in contract from the Ministry of Environment and Climate Change, it undertook a range of studies in order to meet the requirements for a Renewable Energy Approval (REA). Several consultation activities with the local population are required, which the proponent needs to document and report back to the ministry. After NRWC had submitted all required project documents,[11] the Ministry of the Environment opened the proposal for public consultation in a period of 60 days between December 2013 and February 2014. The ministry received 2,572 comments, out of which 2,224 were submitted online and 348 in writing (Government of

Table 5.2 Clustered comments to the proposed NRWC Project

– Impacts on the natural environment;	– Tourism and recreation;
– Impacts on archaeological resources;	– Specific REA reports and studies;
– Consultation undertaken for the project;	aeronautical obstruction;
– Health and safety impacts;	– Local infrastructure;
– Impacts on property values;	– Stray voltage;
– Impacts on birds and bats;	– Setback distances;
– Impacts on geology, energy efficiency and the rising costs of electricity;	– Spacing between turbines;
	– Transmission and collector lines;
– Impacts on the economy;	– Natural disasters/storms;
– Impacts on emergency response;	– Transportation and traffic;
– Shadow/light flicker;	– Television;
– Project location;	– Radio and GPS reception;
– Decommissioning;	– Power failures;
– Noise impacts;	– Possible claims/lawsuits/legal action;
– Waterbodies and hydrology including impacts to drainage;	– Impacts on aggregate resources;
	– Complaints and maintenance.
– Impacts on land use, including impacts to farmland;	
– Visual impacts.	

Source: Government of Ontario 2014.

Ontario 2014). Table 5.2 provides a clustered overview of the submitted concerns published by the ministry.

The ministry's release further stated that there were also comments that requested a moratorium until the Health Canada study was completed. The list reflects the high diversity of concerns that were raised. In November 2014, the ministry granted the REA to NRWC addressed each of the comments (see Government of Ontario 2014) by reference to the fact that all of NRWC's supplied studies were reviewed and approved by the corresponding ministry. For example, the Ministry of Natural Resources and Forestry had approved the required Natural Heritage Assessment and the Ministry of Tourism, Culture and Sport had approved the Archaeological Assessment. Responding to health concerns, the ministry referred to the "scientific evidence" provided by Ontario's Chief Medical Officer of Health. Moreover, the ministry prescribed the formation of a Community Liaison Committee made up of individuals from the community and NRWC. The aim was to "to keep the lines of communication open during the implementation of the Niagara Region Wind Farm, communicate issues that arise during implementation of the project, and report on an annual basis" (Government of Ontario 2014).

The township and regional planners' storylines

Before presenting the storylines of the local anti-wind grassroots organization, the perspectives of two township planners of West Lincoln and Wainfleet and one upper-tier regional planner of Niagara Region are introduced (Anonymous interviews planners Wainfleet (RG), West Lincoln (KS), Niagara Region (AV)). Although the planners are not directly involved in the local dispute over wind turbines, their storylines still reveal valuable insights on the discursive background against which local anti-wind storylines emerge at the local level. Both planners from West Lincoln (KS) and Wainfleet (RG) generally appreciate the decision to move the overall planning authority from the township to the provincial level in wind energy matters, assuming that the township level lacks the resources in terms of staff, capacity and expertise to carry out such projects. Highlighting the severity of the public controversy over wind turbines, RG states that he would probably "not be working for the municipality because there would have been a grassroots call for my head, I'm sure" (Interview, planner Wainfleet RG). Both planners explain that the township level is usually very important in the communication of anticipated developments in the municipality, including wind turbine projects. The township office is an accessible point of information and usually serves as a communication and information platform for the municipalities' residents. The fact that this possibility for local residents is now removed and that they can only acquire information from the developer is regarded as a major driver in public discontent:

> Municipalities are always the first phone call that people have, and when we have no say in the matter, it just increases the level of frustration. We cannot offer any input, any guidance or any support to residents, or just citizens. And the province, even less. They say, "Ok, well, send an email with your comments". We had hundreds of e-mails submitted or comments submitted from Wainfleet and the province basically ignored them, so what is the point.
>
> (Interview, township planner Wainfleet RG)

The quote shows that RG deems the stream of information between the province and the township level as unsatisfactory as he had experienced that many requests for information remained unaddressed. This quote exemplifies also that the superseding of the municipal level is regarded as a major factor in public discontent over wind turbine development, as the municipal level is usually the first point of information for rural residents. The regional planner (AV) explains that the fact that decisions are taken in Toronto is further aggravating the situation:

> In Ontario, there is a real sort of underline undercurrent about Toronto. And that Toronto makes all the decisions for the province. And then they'll say, … it'll be taken in a negative context. If you go somewhere in a small town, some of us, they are all: You're from Toronto. And it's not a compliment. It's meant to be a slight that, you know, you are one of those people that makes all the choices for . . . you know, you live in one percent of the land mass of Ontario but you make all the choices for the province. And so there's that whole idea that the act and the decisions are driven by people from Toronto who have never been to Wainfleet, they have never been to

West Lincoln. But they are going to tell the people in West Lincoln that they know what's best for them. And that these turbines will go there. So that is a big piece of the sentiment.

(Interview, planner Niagara Region AV)

According to AV, the absence of local input and the transfer of decision-making over wind turbines to the provincial level aggravate public disenchantment over wind turbines because there is a strongly negative perspective of "Toronto" in general. This also refers to the feeling that the rural areas in Ontario are deeply different from the urban areas of the Greater Toronto Area, which is also endorsed by the electoral divide of rural and urban Ontario. While the Progressive Conservatives have most of their votes in the rural areas, the Liberals are usually elected to a larger percentage in the urban areas.

The local anti-wind storylines: grassroots organization and councilors

This section presents the storylines of five members of the two main anti-wind grassroots organizations that are active against the wind turbine projects in West Lincoln and Wainfleet. Those are West Lincoln Glanbrook Wind Action Group (WLWAG) and Mothers Against Wind Turbines (MAWT). Both have official legal status as incorporated organizations. All five interviewees taking part in the focus group discussion belonged to WLWAG and three were a member of MAWT at the same time. They wished to stay anonymous and are referred to as HV, CH, YV, SM and OW. Both organizations work closely together and are sometimes difficult to keep apart (Focus group discussion, Ontario). The focus here is on WLWAG, which was formed in 2010. The WLWAG members were not sure about their precise membership, but they reported around seven executives and approximately 300–350 people on their mailing list. The group positively refers to the umbrella organization WCO. Focusgroup participant HV for example calls WCO's president Jane Wilson[12] "our hero" (Focus group discussion, Ontario, HV) and SM states that "these are celebrities in my mind, they are heroes!" (Focus group discussion Ontario, SM). Next to the perception of the local grassroots activists, the viewpoint of two local councilors of West Lincoln and Wainfleet will be presented here (Anonymous interviews, councilor West Lincoln (CWL) and Wainfleet (CWF)). Both are closely connected to WLWAG.

The storylines of the interviewed WLWAG members and the local councilors largely revolve around the perception that there is "true" and "false" information on the wind turbines and that information is "withheld", "ignored" and "suppressed". This frame is recurring in all of the following five aspects of the local anti-wind storylines: their sources of information, their main arguments, their view on the space of participation, group activities and the view on politics. These aspects will now be presented.

Sources of information, forming opinion and a perceived lack in information

The group members first heard about the wind turbines when the developer of the first project announced an open house information event in the newspaper in 2010. At the time, there was already a negative atmosphere against the project among township councilors, as they

> were pretty upset about the whole thing because the Planning Act had been changed and the municipality itself can do nothing in regards to where these turbines are gonna be placed, the siting of them, what roads they can use.
>
> (Focus group discussion, Ontario, CH)

The interviewees affirm that at the time, they did not know much about wind turbines. They attempted to fill this information deficit by internet research. The WLWAG member YW did not care about the first five wind turbines as they were situated far away from her home. Research on the internet that revealed the dimension of the NRWC project made YW change her mind. She acknowledges that she had not known anything about wind turbines before:

> Well and plus I didn't know there was anything bad about it, I had no idea, I just thought well you get a few of these turbines, there's a few by Dunville, and whatever, and then one day my neighbor said that we were gonna get them – this other project out this way, and ah and then we just looked it up on the internet, and the first thing we saw was the map of how many they're putting in in southern Ontario, and that was enough, I thought no, that can't be [...] there's no way anybody can look at that and then think this is OK. And then you start to find out how bad it is, but it was a learning ... I mean we knew nothing, and then you ... you start to learn.
>
> (Focus group discussion, Ontario, anonymous participant YW)

Researching on the internet played a major role in forming her opinion on wind turbines – and not only for YW. HV had researched negative effects on property values and electricity costs and brought them to the open house meeting of the project developer NRWC, but it frustrated her that the answers she was provided did not match her researched information. Furthermore, "they wouldn't really answer any questions, pointed questions like where are these turbines going to be located, how big are they, why are you doing this?" (Focus group discussion, Ontario, HV). SH adds that "it's important to note that the terminology for that is consultation, but it is not a consultation, it is a one-direction presentation" (Focus group discussion, Ontario, SH). The open house meetings were thus experienced as very unsatisfactory with regard to being provided information and asking critical questions.

Similarly to YW, SM was initially not opposed to wind turbines. He actually liked the idea of using an abundant source of power: "wow, to get electricity from the air like this, at no cost to anybody" (Focus group discussion, Ontario, SM). He states that "being a technical person I was really in favor of renewable energy projects but not the way this has been done" (Focus group discussion, Ontario, SM). When he was moving into the area, he noticed anti-wind signs put up on the streets about stopping the wind turbines, so he decided he "better become an expert in this area" (Focus group discussion, Ontario, SM). Participant SM reports to have called the project manager to acquire more information, but he found his answer unsatisfactory. He consequently started to do internet research and found health impact testimonies of people that lived 2.1 km away from turbines. As he was 4.7 km away from the nearest turbines, he "really didn't expect issues" (Focus group discussion, Ontario, SM). Two weeks after the turbines went operational however, SM started to experience adverse health impacts in the form of ringing in his ears, vertigo, dizziness, high blood pressure and he noticed "a very subtle noise and a vibration in my home" (Focus group discussion, Ontario, SM). He reports he suffers from nine out of the 11 potential health symptoms of living next to wind turbines.[13] His family doctor concluded that his symptoms could only be caused by the turbines as no other reason was found. SM concedes that he was generally sensitive to noise and reports to be able to hear bug-repelling devices, but his property is connected to the bedrock which transmits the vibrations from the turbines. Doing more research on the internet, he found that there were many reports from around the world of people with similar symptoms living close to wind turbines. He was impressed by the detailed description and presentation of their cases:

> I found this Quixote's Last Stand website, and how there are just on this one website there are over 2000 wind actions groups worldwide, and I thought to myself these people have other jobs and they're doing this out of their family time, which is a small sliver of time, it must mean a lot to them. So I found that a lot of other people are having the exact same issues I started to have … and for me then, also my daughter started to have ringing ears at age 11.
>
> (Focus group discussion, Ontario, anonymous participant SM)

SM thus concluded that since people invested a lot of time into fighting wind turbines, this must indeed be a serious issue. Existing websites of anti-wind groups thus played an important role in feeling reassured about his concerns and his emerging critical view on wind turbines. Websites from other anti-wind groups were also a major source of information for the township councilor in Wainfleet (CWF). She reports that both online research and communication with anti-wind groups had been important in gathering information:

> I just went online and, different newspapers, I would read articles and people would send things. I mean, once you get hooked up with a anti-wind group, you get inundated with, once they have your email you get sent all sorts of information. So, I mean, and not everything you read is gospel, I get

that. So if I read an article in, back in the beginning, I'd say, "Ok, well, I don't know that I think that's 100% the truth, so I'm gonna go and figure out, go Google (....) So, I mean you go to the meetings and you share information that you have, and they share websites and blog-spaces and all that other kind of stuff.

(Interview, CWF)

This quote shows that anti-wind groups kept sending her a lot of material on wind turbines that the councilor sometimes verified by online research. Joining group meetings further contributed to gathering and sharing information. CWF has now become an outspoken opponent of wind turbine development, although she initially deemed it a good tax base for Wainfleet. She reports that doing research on the internet, she came across testimonies of people leaving their homes due to adverse health impacts from wind turbines. As she is convinced that "you don't walk away from that kind of investment unless there's something drastically wrong" (Interview, CWF), this was the turning point in her opinion towards wind turbines.

Online research and the information disseminated by other groups contributed not only to the questioning of the information provided by the project developer, but also to a skeptical attitude towards information provided by the government. When the Ontario Chief Medical Officer of Health declared that there was no connection between turbines and adverse health impacts, CWF objected: "I thought, no, I'm sorry, you might be a doctor, but you're wrong" (Interview, CWF). In the same vein, SM expressed his concern that the medical profession, the ministry as well as the proponents do not have correct and sufficient information:

So I became concerned when I noticed they don't have a complete picture of the information, so they're doing an experiment on us, kind of like what the Nuremberg code was written to avert.

(Focus group discussion, Ontario, anonymous participant SM)

Comparing the development of wind turbines to illicit medical experiments on humans, this quote exemplifies how seriously the situation is experienced. The members of the group are not only firmly convinced that the government and the developers do not have enough information on the adverse health impacts of wind turbines, they also think that the government and the developer act in outright sneakiness and secrecy, suppressing the "true" information. This contributes to a deeply felt mistrust of developers' and government information. In the same vein, the media is also regarded as biased. SM is concerned that though many people suffer from the same issues, "the media is not reporting those issues, the media is only reporting when wind opposition is having a defeat" (Focus group discussion, Ontario, SM).

The view that the proponents, the government and the media either do not have or suppress the "true" information contributes to a general self-perception of the anti-wind activists that they fight the good fight. The interviewed councilor of West Lincoln (CWL) compares the fight against the wind developments with historical struggles against wrongful policies that were later-on corrected by true information politics:

> History is full of bad ideas that communities went down together, and it was the continuation of information and dissemination of information that eventually changed policies, whether it is the slave trade, or whether it was land purchases, or whether it was many other policies. There is not a country in the world that looks back at an unblemished history and says, "Yeah, every decision we made as a collective was the right decision". There's not- and so, at one time they thought in the Canadian context that taking First Nations people's children out of their homes and putting them in boarding schools was a really good idea and now our newspapers are full of how bad that was as an idea, as a concept. If nobody spoke out, we would still be doing that, but it is such a bad policy and as time wore on, we realize and we come back and we hang our heads with shame at those points in our history. So we continue the fight. And I honestly believe that 25 years from now, we will look back on this shaking our heads, going, "What were they thinking. It was so economically badly, hamstrung.
>
> (Interview, CWL)

CWL is thus convinced that one day, the "true information" will be acknowledged that wind turbines are a bad choice for the economy and compromising peoples' health. The historical examples that CWL provides further highlight that wind turbines are regarded as a severe historical mistake, comparable to massive human rights violations.

Main arguments and view on energy system

Not only health issues figure prominently among the group's arguments against wind turbines. The economic side of wind turbines is also regarded as especially problematic.[14] The first WLWAG group meetings revolved around the issue of whether there would be negative impacts upon land value resulting from the proposed wind turbines. For HV, the fear of a possible devaluation of her property prompted her to sell it even before the wind turbines were set up. In a written statement on the projects, HV writes that "Our municipality and the non-participating receptors are not willing partners in this proposed transformation of West Lincoln from agricultural land use to an industrial wasteland" (HV, no date). This perceived "industrial character" of wind turbines is intrinsically linked to the fear of property devaluation, as "of course people who are saying no, I'm not going to buy your property are considering the aesthetics" (Focus group discussion, Ontario, HV). By referring to wind turbines as "industrial wind turbines", the members of the group use the wording that is also put forward by the umbrella organization WCO.

Next to property devaluation, economic concerns also relate to the rise in electricity costs that the group associates with the increase in renewable energy in Ontario's power mix. The FIT tariff as an overly high "subsidy" is regarded as the core of the problem:

> I would say the companies that are behind these projects are just acting out of their own greed, to make more money, and our government is stupid enough to set the subsidies so high that we say they don't run on wind, they run on subsidy.
>
> (Focus group discussion, Ontario, anonymous participant CH)

This exemplary quote illustrates the group's critical view on subsidies and the "greedy" character of involved companies. HV and others in the group contend that their "escalating" electricity bills are also attributable to the wind turbines, expressing worries that this would deteriorate Ontario's debt situation. Wind turbines might eventually "bankrupt this province" and also "economically bankrupt the people" (Focus group discussion, Ontario, HV) in the communities because of higher electricity costs and property devaluation. CWL agrees with those economic concerns and contends that Ontario already has a surplus in power generation and therefore does not need new capacity. At the same time, he affirms that the top-down approach via which the turbines were implemented is the number one reason to object to the Green Energy Act (Interview, CWL). This will be discussed further down.

Summarizing, the reasons why the group opposes wind turbine development mainly revolve around adverse impacts on human health and economics. They however state that in general, renewable energy can make sense, but in a different way than it has been done. OW is in favor of "individual solutions" per household, and "if some people want the FIT contract for their house, they should be subsidized, not big industry" (Focus group discussion, Ontario, OW). SM highlights the fact that the first chairman of WLWAG had "been living off solar power for over a decade I believe" (Focus group discussion, Ontario, SM). YW agrees that renewable energy can make sense, but only on an individual level:

> And I think really there are parts of renewable energy that completely make sense, like if you wanna put up a solar panel I'm gonna caution you about where you put it because you do not wanna live in an electro-magnetic field, so you don't want it on your bedroom, but if you can put it so it's not gonna impact the household so you put it on your garage or your barn, although maybe someone would tell me I have no concern for the cows, you know, there are sensible ways of doing . . . things . . . but what happened is in those individual projects, there was no money for the big corporations. And so I think that's why we're ended up in this mess.
>
> (Focus group discussion, Ontario, anonymous participant YW)

This quote highlights again that health issues figure prominently not only with regard to wind turbines but also with solar power. The interviewees further state that renewable energy would make more sense in the far northern parts of Ontario where proper setbacks could be implemented that do "not hurt people" (Focus group discussion, Ontario, YW). This further highlights that the group has experienced wind turbine development in their neighborhood as an industrial project of private companies that were regarded as being "generously" subsidized by the government to the detriment of human health and economics.

View on space of participation: lack of democracy

The general perception of having information withheld from them while having researched the "real" information also influences the group's view of the decision-making system for wind turbine planning. Not surprisingly, this space of participation is regarded as inherently undemocratic and unjust. The issues that are of concern to the group, such as property devaluation or low frequency noise, are not addressed, according to HV:

> You have all of these issues which are never addressed – never addressed! – and cannot be addressed because it's not . . . it's not seen as a problem or you're . . . you're just mongering if you bring any of these issues up, but it doesn't negate the issue, I mean to bury your head in the sand does not change the science of the thing.
>
> (Focus group discussion, Ontario, anonymous participant HV)

This quote illustrates that the "science of the thing" is regarded as the non-negotiable truth that exists irrespective of someone denying or ignoring it. "Burying your head in the sand" in terms of negating or hiding away from this "truth" does not eradicate the existence of the "truth". The space of participation, according to the interviewees, did not allow this "truth" to be included into the decision-making process. The interviewees feel that it is instead designed in a way as not to let anyone raise particular questions. This not only relates to the non-consideration of the information they researched, but also to the physical experiences they had with wind turbines. When SM experienced serious health issues, he contacted provincial and local authorities and the proponent. He reports that he was "calmly, objectively submitting my experiences and reporting issues that needed to be reported" (Focus group discussion, Ontario, SM):

> I didn't just say on this day this happened, I also said, you know, I am observing that it's most prevalent, this vibration in my bedroom, when I am downwind from the turbines, so the turbines are facing away from home and the back of the turbine is toward me. And I offered to work with them, full open – nobody has ever replied, nor asked me questions or anything. The only efforts to protect us have been me working with my family doctor, and he's saying I don't know what else it can be from. So this is the start to my concern.
>
> (Focus group discussion, Ontario, anonymous participant SM)

SM thus states that he reported in detail about the circumstances of his adverse health symptoms, but as he did not get any satisfying answers or no response at all, he started to feel helpless as to whom to report his issues to. The group expresses that their concerns about health impacts are not taken seriously. CH reports that they are often told that they experience those issues because they were expecting them, "you know, because like I'm gonna get something because I'm expecting it to happen because I've been fighting this" (Focus group discussion, Ontario, CH). SM adds that people sent him psychosomatic studies that are meant to explain his health issues. For the group however, SM is the proof that these allegations are wrong because SM was in favor of wind turbines before and did not expect any issues. SM also hints at the difficulty or impossibility of proving that one is subject to emissions in the "infrasonic frequency range, which means that you can't hear them" (Focus group discussion, Ontario, SM) and measurement tools are expensive.

Another important component of the Ontario space of participation is the Environmental Review Tribunal (ERT). WLWAG requested two ERT hearings for the two earlier projects and Mothers Against Wind Turbines one against the larger NRWC project. All three ERT cases were dismissed. In the NRWC case, the decision states that "the Appellant did not call an expert witness to give opinion evidence" and therefore, the evidence they provided was "limited" (Environmental Review Tribunal 2015, p. 45). This hints to the main factors of criticism that the group members raise against the ERT procedure: the burden of proof, the smaller weight of lay testimony and the need for costly expert evidence.

The West Lincoln councilor considers the ERT a "complete waste of time" and very "biased", as the tribunal members are appointed "by the same people who put the Green Energy Act" (Interview, CWL). He also regards the fact that the appellants have the onus of proving serious impacts on human health or the environment by expert opinion as "untenable" (Interview, CWL). SM criticizes the fact that no lay testimonies are considered as double standard:

> This ERT process, it is not an integral process, it is a process that is geared to achieve a certain outcome and it uses various tactics such as exclusions of testimony and striking of evidence and experts flown in who may have an expert as a doctor but maybe in a different field, but they seem to be able to say whatever they want when they're on the stand. Yet I, who am living in this situation, who have made these observations, have documented them carefully, have confirmed them, my words mean nothing there.
> (Focus group discussion, Ontario, anonymous participant SM)

SM thus regards it as inherently unfair that his own experiences are not taken into account and the experts' statements by the project developer are deemed more relevant. OW calls the ERT a "kangaroo court" and states that "here you're supposed to be a democracy, but it turns out it's not" (Focus group discussion, Ontario, OW) as they do not stand the same chances as the private companies to deliver expert opinion. If they had the necessary financial resources, the group affirms, they would engage in a judicial review of the wind projects because they are convinced that some of its details are against the law and the government is not adequately enforcing the rules. However, they lack the financial resources: "And you see the wind company has millions, and we do spaghetti dinners, you know?" (Focus group discussion, Ontario, CH).

The fact that the ERT did not change anything left them with a deep feeling of resentment and the impression that the GEA is a piece of legislation that just aims at pushing through the development of wind turbines at the expense of economics, health, and the environment. CWL's opinion on the GEA is in line with that of the group in terms of the rejection of the top-down approach and the feeling of unfairness with regard to the imbalance in financial resources between the residents and the wind companies:

> I think that the Green Energy Act is the most top-down or dictatorial, and don't use those terms lightly, because a lot of people throw that around (...) We have had absolutely no input in terms of siting, sizing, setbacks, bylaws, on every front we have tried to control or manage these wind turbines, and we get shut down, we get threatened with legal action because we're fighting the Green Energy Act, (...) you know our hands are tied. (...) the type of money that's behind the wind companies is just so large and have deep pockets and able to- at least even if they don't win they threaten to bankrupt you in the courts. So that's not necessarily a just system.
>
> (Interview, CWL)

CWL further states that this opinion is not solely his private opinion in council, but that the other councilors share this view on the GEA. He further specifies that "in policies like this, it is very easy for urban centers to propose this into the blue sections" (Interview, CWL), referring to the fact that the GEA stems from the Liberals who are generally more represented in the urban areas. The "blue" areas of Ontario that vote for the Progressive Conservatives however constitute the larger landmass of Ontario. All in all, the fact that there is "no avenue to take our complaints" contributes to a high level of frustration which, as HV assumes, will ultimately result in "vandalism" (Focus group discussion, Ontario, HV).

Group activities

Among the various activities that WLWAG engages in are joining rallies against the proposed projects at the developers' open house meetings, organizing information meetings for the community, collecting signatures for petitions and engaging in official appeals. As a lack of information is regarded as the main hindrance to a broad mobilization of support, disseminating information is among the most important activities of the group. Members of the group gave a number of presentations to the West Lincoln Township Council and Niagara

Regional Council. They called upon both councils to ask the province put a moratorium on wind turbines in place until there was certainty about the results of the Health Canada study. The group also delivers general information to both councils directly. The interviewed councilor in West Lincoln confirms that he stays up to date using the information that the group sends him: "They do not necessarily come to me for guidance in terms of what to do, they keep me more informed" (Interview, CWL).

WLWAG was initially subdivided into several sub-groups who focused on researching specific topics. Comparing the provided information with their own research represents one of their core activities. For instance, the group discovered online that ten other international wind farms employed the same turbine model but used a higher sound power level than the proponent of the proposed project in their predictive noise modelling. The group concluded that using this higher sound power level would result in a considerably higher number of affected residential buildings in their area, as sound power levels impact the required setbacks. The group addressed this in the hearing of the Environmental Review Tribunal, but their concern was dismissed.

The group convened a number of meetings in the community where they focused on spreading information on the planned project and wind turbines in general. One central focus was to spread information on health impacts. A WLWAG media release of a meeting in February 2011 reported on a community meeting with 140 residents in which they displayed video clips of testimonies by individuals living in the vicinity of wind turbines who have experienced negative health effects (WLWAG 2011). Those testimonies clearly disprove, according to the media release, the claim by the provincial government that there is no scientific evidence of negative health effects, environmental impacts or property value loss. At the meeting, it was proposed that if only fifteen percent of the residents of the nearby homes are affected by the reported health impacts, 100 neighbors are "destined to become seriously ill" (WLWAG 2011). In a 2015 public call for donations, they refer to the NRWC project and state that "1160 victims can expect to suffer health impacts" (WLWAG 2015). This shows again that health impacts are one of their main topics to mobilize support.

The lack of a general awareness among the population on the "real facts" also contributes to the difficulty of mobilizing people, according to the group members. The interviewed councilor of Wainfleet states that because people do not have the full information, they are not "connecting the dots because they think it's green. The overall impression is that they're green and the wind and the solar companies, they promote that all the time" (Interview, CWF). If people were truly informed about the workings of wind turbines, it would be easier to mobilize more people to join their struggle. On the other hand, CWF also states that "there's a lot of head in the sand going on" (Interview, CWF), but she acknowledges that many people are busy with their family and cannot spare time to join them. Overall, the group considers themselves as being in the minority, as SM puts it:

There's another class of person who is probably in the majority who is unaware or they may read it but oh I'm busy playing X-Box so I don't wanna research it.

(Focus group discussion, Ontario, anonymous participant SM)

Being unaware of "what is really going on" is thus regarded as a main hindrance for people to get involved in their group. Additionally, many aspects of concern are difficult to decipher. According to the group, people do not connect their symptoms to, for instance, low frequency noise because it is not easily identifiable. In this context, SM tells the story of a nurse that came to see him who reported similar health symptoms as he had, but she had not connected them to the wind turbines.

View on and interaction with politics

SM understands that a private company is not "rushing to shut them off immediately" but "as a tax paying citizen", politics has the mandate to "take care of the health of the community" (Focus group discussion, Ontario, SM), so he expects them to research his issue. Being in close contact with the Township Councils of West Lincoln and Wainfleet, the interviewees express the concern that the municipalities can no longer protect the economic and health of their residents: "They are sorry, they agree with me, but they can't do anything. This is the responsibility of the Province of Ontario and the Ministry of Environment of Ontario" (Focus group discussion, Ontario, SM). Yet, the group asked the municipalities to take action and gave some presentations before West Lincoln council to put a municipal by-law for larger setbacks in place. HV however was disappointed that West Lincoln Council did not adopt a corresponding resolution, in contrast to Wainfleet, but only put it to the minutes after they had checked with a lawyer who doubted the legal feasibility of the by-law. The general relationship between West Lincoln Council and WLWAG is nevertheless positive, as YW summarizes: "the council understands our concerns, and they can do nothing about them, absolutely nothing – 'Too bad, so sad, sucks to be you'. But no one can do anything about it". (Focus group discussion, Ontario, YW). As the municipal level is regarded as not being able to do much anyway, the group's anger is more directed towards the provincial level.

The councilor of Wainfleet states that as it is the province who has decided on superseding the municipality, "that pretty much ties our hands so we're stuck with it for the next four years until the next provincial election" (Focus group discussion, Ontario, CWF). Replacing the government in the next election is therefore regarded as one option to change the situation. The group has a good relationship to its Member in Provincial Parliament (MPP) Tim Hudak, the leader of the Progressive Conservatives from 2009 to 2014. In the run-up of the October 2011 provincial elections, WLWAG openly called to vote for him, believing that "our local MPP Tim Hudak, leader of the Progressive Conservative Party is our best choice politically in putting a stop to the approval and construction of Industrial Wind Turbines in the Niagara Region, as well as across the province" (WLWAG/GWAG 2011). The then leader of the umbrella organization WCO, John Laforet, was also cited in the WLWAG 2011 newsletter as openly supporting Hudak, who "gives voters a clear choice in the fall":

> We need to awake from this 'green dream' and realize the mirage that is the Green Energy Act. The issues and concerns that directly threaten our local communities are numerous. If Hudak were to be elected in the fall election, he has vowed to eliminate anti-democratic elements of the Green Energy Act, and to return decision making abilities to local municipalities – thus working towards the restoration of our democracy and true Canadian heritage.
>
> (WLWAG/GWAG 2011).

This statement shows that WLWAG unconditionally supports Tim Hudak and trusts him to "eliminate anti-democratic elements" of the contested legislation. Even if Tim Hudak was ultimately not successful in the 2011 elections, he continued to support the anti-wind campaign after he had quit the leadership of the PC in 2014 and remained an MPP for Niagara-West-Glanbrook. While WLWAG thus enjoyed unconditional support from their representative in provincial politics, this good relationship did not lead to any change of their situation at the local level.

Notes

1 This has also been confirmed by two focus group discussions with local opposition groups carried out in 2015.
2 Health concerns relate to adverse effects on human well-being, including dizziness, ear ringing, sleep disorder or nausea. Those are often put in context with low frequency noise or infrasound that wind turbines emit. It is debated whether those effects are directly related to wind turbines. A fierce proponent is the US-based nurse Nina Pierpont who published a book on the "Wind Turbine Syndrome" in 2009 that is widely known in Ontario (Pierpont 2009).
3 *The Wind Turbine Syndrome* is a book published by the American nurse Nina Pierpont in 2009 and is widely known in Ontario (Pierpont 2009).
4 Love Canal is a neighborhood in Niagara Falls, New York, adjacent to a hazardous waste dump. A neighbors' initiative started to raise the emerging health concerns publicly in the late 1970s. The case has become one of the early reference cases for the Environmental Justice Movement in the US.
5 Reflecting an historic importance of hydro-electric power in Ontario's energy mix, energy bills are still called "hydro bills" and electricity prices "hydro rates" after the electricity corporation Hydro One.
6 The Ontario Auditor General Bonnie Lysyk revealed in December 2015 that Ontarians had paid CAD 37 billion above market price for electricity over the past eight years (CBC News 2015). The reason why Ontario hydro bills have increased can be attributed to a number of facts, out of which one is the oversupply in power, but also the fact that renewables account for 6.3 percent of electricity generation but for 16.3 percent of electricity costs, which has been attributed to the high initial FIT tariffs of the GEA (CBC News 2016b).

7 The first quoted recommendation to further streamline the regulatory approvals process was directed at smaller solar projects and at accelerating the approvals process for large-scale projects within the regulatory ministries (Ministry of Energy 2012b, pp. 14–15).
8 In the run-up of the 2011 elections, the planned construction of gas plants in traditional Liberal ridings was cancelled. This was publicly criticized as a dishonest strategy to appease local voters who opposed the plants. The cancellation cost of one billion CAD contributed to the controversy.
9 While the FIT for small-scale renewable generation was still carried out in five rounds since it was closed for larger projects in 2013, the program also ended in 2017 as the procurement target for that year was met and no further procurement was anticipated (IESO 2017a; Ministry of Energy 2016a).
10 As council rejected the fund, NRWC ultimately set up another fund for which community projects could apply for funds directly.
11 Those included, amongst others, a Consultation Report, an Environmental Impact Study, a Noise Assessment Report, a Natural Heritage Assessment, a Waterbody Report and an Archeology Report.
12 Jane Wilson became WCO's president after John Laforet stepped down in late 2011.
13 These 11 health symptoms refer to the list of health problems that have been associated with wind turbines by the "wind turbine syndrome" (Pierpont 2009).
14 The focus here is on their main arguments. Next to those, environmental concerns were also raised, but to a lesser degree (e.g. the consumption of valuable agricultural land or carbon footprint of turbine manufacture).

References

Baxter, Jamie; Morzaria, Rakhee; Hirsch, Rachel (2013): A Case-Control Study of Support/Opposition to Wind Turbines: Perceptions of Health Risk, Economic Benefits, and Community Conflict. In *Energy Policy* 61, pp. 931–943.
Bolichowski, Jeff (2012): Welcome to the Green Energy Capital of Canada. In *St. Catharines Standard*, 7/26/2012. Available online at www.stcatharinesstandard.ca/2012/07/26/welcome-to-the-green-energy-capital-of-canada, checked on 12/12/2017.
Bolichowski, Jeff (2013): Region Won't Revisit Wind Vote. In *St. Catherines Standard*, 9/20/2013. Available online at www.stcatharinesstandard.ca/2013/09/20/region-wont-revisit-wind-vote, checked on 5/24/2017.
Bues, Andrea (2018): Planning, Protest, and Contentious Politics. In *disP – The Planning Review* 54 (4), pp. 34–45. DOI: 10.1080/02513625.2018.1562796.
CBC (2015): Wind Rush – Doc Zone. Canadian Broadcasting Corporation. Available online at www.cbc.ca/doczone/episodes/wind-rush, updated on 4/10/2013, checked on 10/5/2016.
CBC News (2015): Ontario Auditor General Blames Lack of Oversight Blamed for High Hydro Costs. Province Paying for Decisions that Were Not Vetted by Energy Board, Bonny Lysyk Says. Canadian Broadcasting Corporation, 12/3/2015. Available online at www.cbc.ca/news/canada/toronto/programs/metromorning/ontario-auditor-general-electricity-1.3348569, checked on 8/15/2017.
CBC News (2016a): Ontario Cancels Plans for More Green Energy Citing Strong Electricity Supply. Canadian Broadcasting Corporation, 9/27/2016. Available online at www.cbc.ca/news/canada/toronto/ontario-electricity-plans-1.3780440, checked on 11/28/2017.
CBC News (2016b): Why Your Hydro Bill is So High, and Why It'll be Hard for Premier to Cut It. Canadian Broadcasting Corporation, 11/22/2016. Available online at www.cbc.ca/news/canada/toronto/ontario-hydro-bills-1.3860314, checked on 8/15/2017.

CMOH (2010): The Potential Health Impact of Wind Turbines. Chief Medical Officer of Health (CMOH) Report. Available online at www.health.gov.on.ca/en/common/ministry/publications/reports/wind_turbine/wind_turbine.pdf, checked on 10/5/2016.

Council of Ontario Universities (2010): ORC – Renewable Energy Technologies and Health. Available online at www.orc-reth.uwaterloo.ca/, checked on 10/5/2016.

Cross, William P.; Malloy, Jonathan; Small, Tamara A.; Stephenson, Laura Beth (2015): Fighting for Votes. Parties, the Media, and Voters in an Ontario Election. Vancouver, Toronto: UBC Press.

Dakin, Dan (2013): West Lincoln Turns Down Wind Turbine Money. In *Welland Tribune*, 5/29/2013. Available online at www.wellandtribune.ca/2013/05/29/west-lincoln-turns-down-wind-turbine-money, checked on 5/28/2015.

Deignan, Benjamin; Harvey, Erin; Hoffman-Goetz, Laurie (2013): Fright Factors About Wind Turbines and Health in Ontario Newspapers before and after the Green Energy Act. In *Health, Risk & Society* 15 (3), pp. 234–250. DOI: 10.1080/13698575. 2013.776015.

Environmental Defence (2016): New poll shows Ontarians strongly support the province's green energy policies. Available online at https://environmentaldefence. ca/2016/05/25/new-poll-shows-ontarians-strongly-support-provinces-green-energy-policies/, checked on 12/5/2017.

Environmental Review Tribunal (2015): Case No. 14–096 Mothers Against Wind Turbines Inc. v. Ontario (Environment and Climate Change). Proceeding commenced under section 142.1(2) of the Environmental Protection Act, R.S.O. 1990, c.E.19, as amended. Environmental Review Tribunal. Available online at www.ert.gov.on.ca/english/hearings/index.htm, checked on 6/20/2015.

Government of Canada (2014): Wind Turbine Noise and Health Study: Summary of Results. Government of Canada. Available online at www.canada.ca/en/health-canada/services/environmental-workplace-health/noise/wind-turbine-noise/wind-turbine-noise-health-study-summary-results.html, updated on 11/24/2017, checked on 11/24/2017.

Government of Ontario (2014): Environmental Registry: Decision on the Niagara Region Wind Corporation proposal. Available online at www.ebr.gov.on.ca/ERS-WEB-External/displaynoticecontent.do?noticeId=MTIxMTM5&statusId=MTg2MDY0, checked on 6/22/2015.

Hill, Stephen; Knott, James (2010): Too Close for Comfort: Social Controversies Surrounding Wind Farm Noise Setback Policies in Ontario. In *Renewable Energy Law & Policy Review* (153), pp. 153–168.

IESO (2017a): 50 MW Procurement Target Reached – December 1, 2017. Independent Electricity System Operator. Available online at www.ieso.ca/en/get-involved/microfit/news-bi-weekly-reports/50-mw-procurement-target-reached--december-1-2017, checked on 12/5/2017.

IESO (2017b): Energy Procurement Programs and Contracts. Large Renewable Procurement. Independent Electricity System Operator. Available online at www.ieso.ca/sector-participants/energy-procurement-programs-and-contracts/large-renewable-procurement, checked on 12/5/2017.

Ipsos Reid (2010): Wind Energy in Ontario. Available online at www.canwea.ca/pdf/ipsosreid ontariosurvey.pdf, checked on 8/20/2015.

Jones, Allison (2017): Ontario Signals Offshore Wind Moratorium will Continue for Several More Years. In *Toronto Star*, 2/13/2017. Available online at www.thestar.com/news/canada/2017/02/13/ontario-signals-offshore-wind-moratorium-will-continue-for-several-more-years.html, checked on 12/13/2017.

King, Arlene (2009): Memorandum RE: Wind Turbines. Ministry of Health and Long-Term Care, October 2009. Available online at www.wind-works.org/cms/fileadmin/user_upload/Files/Wind_Turbines_Health_Dr_Arlene_King_oct_09.pdf, checked on 10/5/2016.

Lees, Janet (2011): Ontario's Giant Slayer. One man's "David and Goliath" Battle to Topple Big Government and Big Wind. In *On The Bay Magazine* 2011, (Summer 2011), pp. 16–19. Available online at http://issuu.com/onthebaymagazine/docs/otb-summer2011, checked on 11/5/2015.

Martin, Chip (2013): Wynne Supports More Municipal Autonomy on Green Energy Projects. In *The London Free Press*, 1/20/2013. Available online at www.lfpress.com/2013/01/20/wynne-supports-more-municipal-autonomy-on-green-energy-projects, checked on 12/6/2016.

Miner, John (2016): The Mainstreet Research Survey Suggests an Even Split in Public Opinion about Ontario's Embrace of Wind Energy. In *The London Free Press*, 2016. Available online at www.lfpress.com/2016/06/07/the-mainstreet-research-survey-suggests-an-even-split-in-public-opinion-about-ontarios-embrace-of-wind-energy, checked on 12/5/2017.

Ministry of Energy (2012a): Moving Clean Energy Forward, Creating Jobs. McGuinty Government Takes Steps to Ensure Successful Renewable Energy Program. Available online at http://news.ontario.ca/mei/en/2012/03/moving-clean-energy-forward-creating-jobs.html, checked on 8/11/2015.

Ministry of Energy (2012b): Ontario's Feed-in Tariff Program. Two-Year Review Report.

Ministry of Energy (7/11/2012): Directive to the OPA: Feed-in Tariff Program Launch.

Ministry of Energy (2013a): New Ontario Government Strengthens Energy Planning. News Release. Available online at https://news.ontario.ca/mei/en/2013/05/new-ontario-government-strengthens-energy-planning.html, checked on 4/15/2016.

Ministry of Energy (2013b): Ontario Working with Communities to Secure Clean Energy Future. Available online at https://news.ontario.ca/mei/en/2013/05/ontario-working-with-communities-to-secure-clean-energy-future.html, checked on 4/15/2016.

Ministry of Energy (2013c): Directive: Renewable Energy Program. June 12, 2013. Ministry of Energy Office of the Minister.

Ministry of Energy (2016a): NUGs, OEFC, FIT, 2015–2020 Conservation First Framework, and the Industrial Accelerator Program. Directive to the Independent Electricity System Operator. Available online at www.ieso.ca/-/media/files/ieso/document-library/ministerial-directives/2016/directive-nug-20161216.pdf?la=en.

Ministry of Energy (2016b): Ontario Suspends Large Renewable Energy Procurement. News Release. Available online at https://news.ontario.ca/mei/en/2016/09/ontario-suspends-large-renewable-energy-procurement.html, checked on 12/2/2016.

Moore, Amanda (2013): West Lincoln Not a Willing Host to Turbines. In *Grimsby Lincoln News*, 5/16/2013. Available online at www.niagarathisweek.com/news-story/3272903-west-lincoln-not-a-willing-host-to-turbines/, checked on 6/9/2015.

Niagara This Week (2011): Wainfleet Council Favours Turbine Moratorium. In *Niagara this Week* 2011, 3/9/2011. Available online at www.niagarathisweek.com/news-story/3312150-wainfleet-council-favours-turbine-moratorium/, checked on 12/12/2017.

NRWC (2011): Notice of a Proposal. by Niagara Region Wind Corporation to Engage in a Renewable Energy Project. Available online at https://lincoln.civicweb.net/document/45240/NRWC_Notice_06July11.pdf?handle=716A1AEF731741AFAB9934 8FF7D8024C, checked on 5/22/2017.

OFA (2012): OFA Position statement on Industrial Wind Turbines. Available online at https://ofa.on.ca/uploads/userfiles/files/OFA%20position%20statement%20on%20 Industrial%20Wind%20Turbines.pdf.

Ontario Ministry of Energy (2013): Achieving Balance – Ontario's Long-Term Energy Plan. Queen's Printer for Ontario. Toronto.

Ontario Ministry of Energy (2017): Delivering Fairness and Choice. 2017 Long-Term Energy Plan. Queen's Printer for Ontario. Toronto.

PC Caucus (2012): Paths to Prosperity. Affordable Energy. An Ontario PC Caucus White Paper.

Pierpont, Nina (2009): Wind Turbine Syndrome. A Report on a Natural Experiment. Santa Fe, NM: K-Selected Books.

Songsore, Emmanuel; Buzzelli, Michael (2014): Social Responses to Wind Energy Development in Ontario: The Influence of Health Risk Perceptions and Associated Concerns. In *Energy Policy* 69, pp. 285–296. DOI: 10.1016/j.enpol.2014.01.048.

Statistics Canada (2011a): Focus on Geography Series, 2011 Census – Census subdivision of Wainfleet, TP (Ontario). Available online at https://www12.statcan.gc.ca/ census-recensement/2011/as-sa/fogs-spg/Facts-csd-eng.cfm?LANG=Eng&GK=CSD& GC=3526014, checked on 11/3/2015.

Statistics Canada (2011b): Focus on Geography Series, 2011 Census – Census subdivision of West Lincoln, TP (Ontario). Available online at https://www12.statcan.gc.ca/ census-recensement/2011/as-sa/fogs-spg/Facts-csd-eng.cfm?LANG=Eng&GK= CSD&GC=3526021, checked on 11/3/2015.

Stokes, Leah C. (2013): The Politics of Renewable Energy Policies: The Case of Feed-In Tariffs in Ontario, Canada. In *Energy Policy* 56, pp. 490–500. DOI: 10.1016/j. enpol.2013.01.009.

Stokes, Leah C. (2016): Electoral Backlash against Climate Policy: A Natural Experiment on Retrospective Voting and Local Resistance to Public Policy. In *American Journal of Political Science* 60 (4), pp. 958–974. DOI: 10.1111/ajps.12220.

Taylor, Susan (2011): Ontario Puts Moratorium on Offshore Wind Projects. In *Reuters*, 2/14/2011. Available online at www.reuters.com/article/us-ontario-wind/ontario-puts-moratorium-on-offshore-wind-projects-idUSTRE71A6Y020110214, checked on 12/13/2017.

TVO (2014): Big Wind. Available online at http://tvo.org/video/documentaries/big-wind, checked on 10/5/2016.

TVO (2015): Wind Power, Wind Problems? | Mar 31, 2015 | TVO.org. Available online at http://tvo.org/transcript/2289438/video/programs/the-agenda-with-steve-paikin/wind-power-wind-problems, checked on 10/5/2016.

Unwilling Hosts (2015): Ontario Unwilling Hosts. Available online at http://ontario-unwilling-hosts.org/, updated on 8/13/2015, checked on 12/10/2015.

Walker, Chad; Baxter, Jamie (2017): Procedural Justice in Canadian Wind Energy Development. A Comparison of Community-Based and Technocratic Siting Processes. In *Energy Research & Social Science* 29, pp. 160–169. DOI: 10.1016/j.erss.2017.05.016.

Walker, Chad; Baxter, Jamie; Ouellette, Danielle (2014a): Adding insult to injury: The development of psychosocial stress in Ontario wind turbine communities. In *Social Science & Medicine (1982)*. DOI: 10.1016/j.socscimed.2014.07.067.

Walker, Chad; Baxter, Jamie; Ouellette, Danielle (2014b): Beyond Rhetoric to Understanding Determinants of Wind Turbine Support and Conflict in Two Ontario, Canada Communities. In *Environment and Planning* 46 (3), pp. 730–745. DOI: 10.1068/ a130004p.

Walker, Chad; Stephenson, Laura; Baxter, Jamie (2018): 'His Main Platform is "Stop the Turbines"'. Political Discourse, Partisanship and Local Responses to Wind Energy in Canada. In *Energy Policy* 123, pp. 670–681. DOI: 10.1016/j.enpol.2018.08.046.

Warren, Michael (2013): How Ontario's Liberals Can Win Back Rural Ontario. In *Toronto Star*, 6/2/2013. Available online at www.thestar.com/opinion/commentary/2013/06/02/how_ontarios_liberals_can_win_back_rural_ontario.html, checked on 8/15/2017.

Wind Concerns Ontario (2014): Response to Health Canada's Wind Turbine Noise and Health Study.

Wind Concerns Ontario (2015): About Us. Available online at www.windconcerns ontario.ca/about-us/, checked on 10/21/2015.

WLWAG (2011): Wind Action Group Going Legal. Media Release. Smithville, ON (February 16,2011). Available online at www.wlwag.com/uploads/5/2/9/6/5296281/wlwag_media_release_feb_16_2010.pdf, checked on 3/11/2015.

WLWAG (2015): West Lincoln Appeal Fund Against Industrial Wind Turbines. Available online at www.wlwag.com/uploads/5/2/9/6/5296281/west_lincoln_appeal_fund_against_iwts.pdf.

WLWAG/GWAG (2011): Wind Newsletter. October 2011. West Lincoln/Glanbrook Wind Action Group. Available online at www.wlwag.com/uploads/5/2/9/6/5296281/october_newsletter.pdf, checked on 5/23/2017.

6 Contention in context

Governmental response to social movements

Wind turbines have become a hot-button issue in many jurisdictions that aim at transforming their energy systems. Both the German federal state of Brandenburg and the Canadian Province of Ontario have tabled strong policies for wind energy development which paved the way for becoming regional forerunners. Both sub-national governments have experienced a backlash in terms of a strong anti-wind movement to which they have responded differently. Brandenburg has kept its ambitions to develop wind energy, while Ontario's program has ultimately been stopped. This study argues that different government responses can be traced to different discursive energy spaces. Those provide the anti-wind movements with diverging capabilities to scale-up their concerns to the sub-national policy-making level. This chapter first compares the characteristics of the two anti-wind movements and the response they received by their governments and then shows how precisely the concept of discursive energy space can help explain different government responses.

Comparing anti-wind movements and government responses

The concept of discursive energy space understands disputes over wind turbines as "struggle for discursive hegemony" (Hajer 1995, p. 59). Several discourses on energy transitions were identified earlier: the *fossil discourse, economic rationalism, ecological modernization,* and *democratic pragmatism.* In both Brandenburg and Ontario, social movements had played an important role in weakening the *fossil discourse* and in advocating for renewable energy, although this only applies indirectly to Brandenburg. The *fossil discourse* is based on a belief in abundant resources and unlimited growth and does not question the use of fossil-fuel resources. The federal Renewable Energy Sources Act (EEG) was largely a product of the German anti-nuclear movement and Ontario's coal phase-out had also been the result of intense lobbying by civil society actors. In the same vein, the Green Energy Act (GEA) Alliance united a broad range of actors who pushed for Ontario's GEA. In Brandenburg, by contrast, the push to promote renewable energy was largely a governmental initiative which followed the discourse of *ecological modernization.* The discourse of *democratic pragmatism,* which endorses community-based renewable energy schemes, was not a central driver behind Brandenburg's renewable energy policies. This is a major factor in why the emphasis on community wind energy schemes has been less strongly pronounced in Brandenburg than in Ontario. Alongside *ecological modernization, democratic pragmatism* was also an important discourse that prompted renewable policy development in Ontario. This discourse was very strong during the initial phase of the GEA and put forward by civil society organizations united under the GEA Alliance. Nonetheless, the characteristics of wind power development in Ontario and Brandenburg turned out similar.

Under the microFIT program of the GEA, more than 26,000 contracts for small-scale solar power and biogas representing over 230 MW of renewable capacity were awarded (IESO 2017), which included a large number of community projects. On the whole, however, wind power has taken the shape of large-scale projects often owned by foreign investors. One exception is a cooperative project in Oxford County which became operational at the end of 2016. This face of wind turbine development characterized by large-scale non-community projects is similar in Brandenburg. While there are a few model projects with community involvement, cooperatively-owned or community wind projects are few and far between. Considering how wind turbines were developed in Brandenburg and Ontario, the opposition against wind energy in Brandenburg and Ontario is unsurprising. As the literature on wind energy conflicts suggests, lack of community-based initiatives and perceived injustices are a key factor behind low rates of acceptance (Goedkoop and Devine-Wright 2016; Gross 2007; MacArthur 2016; Wolsink 2007).

Comparing collective action frames and movement organization

The anti-wind movements in Brandenburg and Ontario fit the definition of social movements as groups that engage in conflictual collective action to bring about social change, form dense informal networks and share a collective identity which goes beyond single events and initiatives (Della Porta and Diani 2006, pp. 20–22). First of all, the anti-wind movements in both jurisdictions engaged in conflictual collective action in order to oppose wind turbine development in their sub-national jurisdictions. This referred to protesting against locally proposed wind turbine projects as well as tackling the sub-national policy-making level to stop or alter the jurisdiction's wind turbine policies. Second, their organization was based on informal networks between local anti-wind groups and their umbrella organizations. Third, they shared the belief that their sub-national government's wind energy policies needed to be changed. Their shared identity related to their feeling of having to bear the burden of the government's renewable energy policies and for some individuals, this also referred to the feeling of being victimized.

Comparing internal organization and strategy

Brandenburg's and Ontario's anti-wind movements correspond to the three elementary types of movement structures distinguished by Rucht (1996), including the *grassroots model*, the *interest-group model* and the *party-oriented model*. The grassroots model correlates to the local anti-wind groups which were characterized by a "relatively loose, informal, and decentralized structure" and "an emphasis on unruly, radical protest politics" (Rucht 1996, p. 188). The local grassroots organizations in Brandenburg and Ontario investigated for this study were indeed organized in relatively loose structures. They were organized and run in a similar manner. Both the Temnitz group in Brandenburg and WLWAG in Ontario did not keep a membership list but both reported about seven to ten active members in their steering committees (Focus group discussions, Brandenburg and Ontario). Both local anti-wind groups engaged in similar activities, including the collection of signatures, disseminating information by organizing information events and generally participating in the available spaces of participation. They focused their attention on the struggle against proposed or existing wind turbines in their neighborhoods by targeting the regional planning processes which determined the wind suitability of areas (Temnitz region) or fighting the proposed wind turbine projects in their area (Niagara Region). The groups' activities were mainly confined to the local level and neither of the two groups achieved their goal of halting the local wind planning process in their area. Another analogy is that both local grassroots organizations considered themselves as the "minority" in their villages and both reported to have difficulties in mobilizing others to join their group. While the Brandenburg anti-wind group speculated that this could have to do with people's general laziness and the legacy of the GDR which suppressed social movements and mobilization, the Ontario anti-wind group traced this to an overall information deficit on wind turbines.

The umbrella anti-wind organizations in Brandenburg and Ontario were typical instances of the interest-group model which is "characterized by an emphasis on influencing policies (...) and a reliance on formal organizations" (Rucht 1996, p. 188), but they also had traits of the party-oriented model. This type of movement structure has an "emphasis on the electoral process, party politics, and, as well, a reliance on formal organization" (Rucht 1996, p. 188). According to the organizations' information, the umbrella organization *Rettet Brandenburg* had around 100 member organizations (as of 2016, interview Thomas Jacob) and Wind Concerns Ontario (WCO) 49 member organizations (as of November 2015) (Wind Concerns Ontario 2015). While both umbrella organizations targeted the political process at the sub-national level with the aim to stop further wind turbine development in their jurisdiction, the anti-wind campaign in Ontario was much more strategically run than in Brandenburg. The campaign of WCO for example targeted the 2011 provincial elections. The organization sought support by drawing on the collective action frame of injustice and by drawing on the strategic framing of "rights". As WCO's president at the time had been a politically experienced community organizer and also experienced with the media, the campaign was embedded in a comprehensive media strategy (Interview, John Laforet). The main goal was to join forces with different actors including local anti-wind groups, local municipalities and politicians from different levels of government. The strategic frame of "rights" allowed for a consensus on the overall goal to fight the GEA as a piece of legislation despite possible differences in opinions about wind turbines. Their most important ally became the Progressive Conservatives who made wind turbines a central topic of their electoral campaign in 2011, calling for a moratorium of wind turbines until health studies are completed and questioning the costs of the FIT. *Rettet Brandenburg*, by contrast, was not run in a similarly strategic way when it came to garnering sympathy and political support. Their campaign did not target an election but took the shape of two public initiatives and a public referendum. They focused on the two central demands: a setback of "10h" and no wind turbines in forests. In contrast to Ontario, the Brandenburg campaign did not follow a visible media strategy nor were they strategically focusing on particular issues to gain the support of municipalities or politicians (Interview, Thomas Jacob).

The comparison of the anti-wind umbrella organizations shows a difference in their strategic organization and the management of their campaigns. Yet, these differences do not automatically explain the observed differences in government response. While a more strategically run campaign certainly contributes to more pressure on a government, the question still arises as to why the different framings of the campaigns led to different pressures on the government. As the president of WCO assumed, the movement would have been much less successful if they had focused the debate on other issues than "rights" and they consciously avoided the framing of "coal vs. wind" (Interview, John Laforet).

Comparing collective action frames

Social movements use collective action frames which inspire and legitimate social movement activities (Snow and Benford 1992). The anti-wind discourses in Brandenburg and Ontario revolved around the collective action frame of injustice. More precisely, anti-wind groups raised arguments relating to environmental justice, urban/rural justice and procedural justice. All three types of arguments were used by the anti-wind umbrella organizations, the local anti-wind grassroots groups and local councilors.

Environmental justice arguments address a perceived "unequal distribution of environmental and economic costs and benefits" (Cowell et al. 2011, p. 554). This kind of argument was both raised at the umbrella and the grassroots part of the anti-wind movements. Both leaders of the anti-wind umbrella organizations in Brandenburg and Ontario drew on environmental justice arguments to support their negative stance towards wind energy development by either regarding them as a further proof of the discrimination or outright exploitation of Eastern Germany (Brandenburg) or comparing them to a toxic waste dump (Ontario)

(Interviews, Thomas Jacob and John Laforet). The speaker of the Brandenburg umbrella organization Thomas Jacob referred to the spatially uneven "burden" of wind power development in Germany which disproportionally affected Brandenburg, leading to higher electricity prices and deteriorating living conditions (Interview, Thomas Jacob). He perceived the forerunner status of Brandenburg as unjust because in his perspective, other federal states in Germany were not doing enough themselves while Brandenburg would sacrifice its scenic landscape. He situated the conflict over wind turbines within a larger conflict between Western and Eastern Germany by sarcastically calling Eastern Germany the "waste dump of the republic" (Thomas Jacob). In a similar vein, the leader of the Ontario anti-wind umbrella organization John Laforet used the environmental justice argument with regard to the potential health issues to which people in the vicinity of wind turbines would be exposed (Interview, John Laforet). He also reported that in his early anti-wind grassroots group in a Toronto neighborhood, it was common to compare the proposed wind power project to the case of Love Canal where a group of neighbors successfully raised awareness around health concerns of an adjacent toxic waste dump in the late 1970s. Similar to the speakers of the umbrella organizations, both local grassroots groups raised environmental justice arguments. Both the local group in Temnitz and the local group in the Niagara Region considered themselves under a severe health threat regarding wind turbine development in their areas and did not feel taken seriously by local authorities (Focus group discussions with local anti-wind groups). Within the Ontario group, the feeling of injustice regarding the economic burden was also a major point of concern. The group regarded it as unfair that they should shoulder the economic burden in terms of property devaluation and rising electricity costs while the private developers were making a profit.

The environmental justice argument can be specified further by the *urban/ rural justice* argument. It highlights the spatial dimension of environmental justice and relates to the view that the rural areas are particularly disadvantaged to the favor of the urban areas. For the local anti-wind grassroots group in Brandenburg, the overarching argument was a sense of injustice that Brandenburg and especially their respective area of Temnitz already had enough wind turbines and would thus sufficiently contribute to the German energy transition, i.e. the *Energiewende* (Focus group discussion, Brandenburg). The group regarded it as unfair that Brandenburg had to sacrifice its beautiful landscape and the well-being of its residents by a "disproportionate" generation of renewable power beyond the region's own energy needs. The speaker of *Rettet Brandenburg* and the anti-wind group in Temnitz region argued that they had nothing against single turbines in their area but became concerned when the number of turbines increased significantly (Interview, Thomas Jacob and focus group discussion, Brandenburg). The interviewees considered it unfair that the rural areas should become the electricity producers for the urban areas while those would not contribute anything to the *Energiewende* (Focus group discussion, Brandenburg). This reasoning highlights the importance of the *Energiewende* discourse in Germany which will be further discussed later.

In Ontario, the local anti-wind group in the Niagara Region considered health and economic impacts as most important, while landscape effects were not articulated to the same degree as in Brandenburg (Focus group discussion, Ontario). Despite the absence of a commonly acknowledged discourse on the need for energy transitions such as the German *Energiewende*, the urban/rural injustice argument was raised by the interviewed councilors and township planners (Interviews with Wainfleet and West Lincoln township planners). They claimed that the superseding of municipalities was in line with a general strategy of "Toronto imposing decisions" upon the rural areas. This refers to the implementation of the GEA which shifted local authority over wind turbine planning from the municipal level to the provincial level. The perspective that the provincial level decides over the rural areas is reinforced by the fact that disputes over wind turbines in Ontario unfolded against the backdrop of a major urban-rural electoral divide. In the two provincial elections preceding the 2018 election, the Liberals as founders of the GEA were elected in most urban areas, including the greater Toronto area with the largest share of population in Ontario, while the rural ridings are mainly represented by the Progressive Conservatives. The New Democratic Party is present in both, but only to a small percentage.

Next to environmental and rural/urban justice arguments, *procedural justice* arguments could also be identified. Those relate to the way in which the decision-making system for wind turbine siting is perceived and encompasses notions of procedural fairness and fair outcomes (Gross 2007). It thus mainly relates to how the space of participation is perceived. The process of wind turbine siting was described as not offering the opportunity for real participation (Focus group discussions, Brandenburg and Ontario). The Brandenburg interviewees complained about a lack in possibilities to direct their health concerns and to have a say in the regional planning process. Among the points of criticism were the disproportionately small representation of smaller municipalities in the regional assembly, the complexity of the process which made it difficult for the average citizen to understand how the process worked, and the reality that their expressions of concern in the context of the formal participation process did not lead to any change. Opponents in the local case study in Ontario complained both about a lack of information and the way in which they could (not) forward their concerns (Focus group discussion, Ontario). In their view, both developers and the government should have provided them with more information. They were also deeply frustrated about their appeal hearings before the Environmental Review Tribunal (ERT), which were both dismissed.

In summary, the overall collective justice frames in Brandenburg and Ontario both focused on injustice, but also included a variety of other concerns such as effects on the landscape, economic impacts or health impacts. One major difference between Brandenburg and Ontario was the demands that were scaled-up to the provincial level and which the government was directly confronted with. Those included the demand to implement a setback of "10h" and no more wind turbines in forests in Brandenburg, and the three demands of re-establishing local planning power ("rights debate"), a moratorium on wind turbines until health studies are completed, and tackling the cost issue in Ontario.

Comparing government responses

How did the Brandenburg and Ontario governments respond to the demands of the anti-wind movements? The focus is on the sub-national government reaction to the demands of the anti-wind umbrella organizations.[1] Investigating government responses, it is important to differentiate between direct and indirect responses. The Brandenburg government was legally obliged to directly address the movement's demands because they were organized as a public initiative and public referendum. The Ontario government, by contrast, did not have to directly address the movement's demands because they were expressed in a general way and not in a specific legal process as in Brandenburg. When governments do not have to address the movement's demands directly, it is generally easier for them to change their policies in response to the movement's concerns without attributing these changes directly to the movement's demands.

The Brandenburg government's response to anti-wind protest can be classified as partly *diplomatic* and partly *exclusionary* (see Table 6.1). The former means that the movement is recognized as legitimate, but their demands are not addressed, while the latter means that government neither regards the movement as legitimate nor adopts its concerns (see Chapter 2). In 2015, the popular initiative of *Rettet Brandenburg* submitted 33,000 signatures to state parliament presenting the demands to require a "10h" setback of wind turbines from the next residential building and to forbid the building of wind turbines

Table 6.1 Government response to anti-wind protest in Brandenburg (BB) and Ontario (ON)

		Accepting the movement	
		Recognizing movement as legitimate	*Denying legitimacy of the movement*
Taking over the movement's demands	*Adopting demands*	Integrative Response (ON): High responsiveness: cost and participation issues were addressed in major policy changes.	Shadow Response
	Declining demands	Diplomatic Response (BB): Information strategy as main response strategy, but the two central demands are not adopted. Reference to regional planning and federal level.	Exclusionary Response (BB): Legitimacy of the movement is partly questioned in light of the need for the *Energiewende*.

Source: Author's Representation Drawing on Gamson (1975) and Koopmans and Kriesi (1995).

in forests. The state parliament adopted a resolution to reject both demands and called upon the state government to reconsider Brandenburg's area target for wind energy and support the regional planning authorities in the designation of regional plans (Landtag Brandenburg 2015). Endorsing the regional planning level reflects the general position of the Brandenburg government. The regional planning level is considered as being the best suited to finding locally adapted solutions instead of implementing standardized solutions at the state level (SPD Brandenburg and DIE LINKE Brandenburg 2014; Landtag Brandenburg 2015).

The *diplomatic* part of the government response refers to the observation that the Brandenburg government did take anti-wind sentiments seriously by engaging in a dialogue and information strategy. This was also reflected in Brandenburg's Energy Strategy 2030 (MWE 2012). The *Energiewende* was regarded as an important situation of societal change in which the local level needs support (Interview, Ralf Christoffers). It was acknowledged that transformation is not easy and that worries emerge. This was tackled by an information strategy which focused on explaining the embeddedness of Brandenburg's energy policy into the federal context and also the role of Brandenburg as an energy export state (*Energieland*) for Germany's security of supply (Interview, Ralf Christoffers). Even if the two demands of "10h" and no wind turbines in forests were declined, they were taken seriously.

Furthermore, the government identified the higher electricity costs for Brandenburg residents as compared to other areas in Germany as a major driver behind public opposition. As the cost structure falls within the responsibility of the federal level, the government called upon the federal level to find a better system of cost sharing and endorsed the changes to the EEG which were believed to tackle the cost question. While this appeal to the federal level shows that the Brandenburg government acted on public opposition against wind turbines, it must also be noted that rising electricity costs were not among the major points of concern of the anti-wind movement (Focus group discussion, Brandenburg). The interviewed speaker of the umbrella organization mainly referred to environmental justice arguments, and the local anti-wind group regarded landscape and health issues as more important (Focus group discussion, Brandenburg; Interview, Thomas Jacob).

The *diplomatic* part of the Brandenburg government's response was thus to acknowledge the movement while still emphasizing the responsibility of deciding over setbacks and wind turbines in forests for regional planning and the federal level's responsibility to tackle the cost issue. This response can also be partly classified as *exclusionary* because it did not regard the movement as completely legitimate. As the then Minister of Energy expressed, the *Energiewende* cannot be realized if every state acts for itself (Interview, Ralf Christoffers), and another ministry representatives stated that if the government yielded to the movement's demands, they would even demand more (Anonymous interview ministry representative). These statements show that the *Energiewende* is regarded as the overall goal for Brandenburg's wind energy development. A focus that is fixated on just local and regional aspects would inhibit the *Energie-wende's* realization. Such a viewpoint indirectly delegitimizes the anti-wind movement which primarily focuses on such local and regional priorities. The second statement also has an *exclusionary* tendency as it does not regard the movement as a serious counterpart with whom to engage in fruitful discussion and reach consensus-based solutions.

In contrast to the situation in Brandenburg, the Ontario government's response to anti-wind protest can be classified as mainly *integrative*. This means that the government both acknowledges the movement as legitimate and adopts their particular concerns. As mentioned earlier, Ontario was not legally obliged to directly respond to the movement's demands; its response was indirect. While the Ontario government did not adopt the movement's demand for a moratorium on the further development of wind turbines until health studies are completed, the policy changes the Ontario government undertook ultimately ended the renewable energy program. The policies thus had the same consequence as a moratorium would have had. The end to the renewable energy program however, happened only in 2016 and before this, the government had already responded to the movement's demands to re-establish local planning authority ("rights") and control costs. Early alterations were the initial lowering of the FIT prices and the 2011 moratorium on off-shore wind turbines. Officially motivated by a further need for environmental studies, this decision can as well be interpreted as a direct response to protest against offshore wind energy, most notably the protest at the Scarborough Bluffs which were one of the first anti-wind protests in Ontario. A more decisive change to the overall policy occurred in 2013. The provincial government suspended the FIT for projects larger than 500 kW and implemented a points system that ranked projects with local municipal support higher in the approvals process. The previous procurement targets for solar, wind and biogas were held on to and the period for their achievement was prolonged (Ontario Ministry of Energy 2013). The renewable energy program for large-scale projects was ultimately ended in 2016 and the 2017 Long Term Energy Plan did not spell out new procurement targets for wind, solar or biogas (Ontario Ministry of Energy 2017).

The policy documents that announced these changes to Ontario's renewable energy policy showed full recognition of the anti-wind movement's demands concerning "rights" and "costs", even if all policy changes were officially informed by and based on large public consultation processes and not directly addressed the anti-wind movement. This includes for example the consultation process for the FIT review and for the Long Term Energy Plan. The demands of the anti-wind movement regarding "rights" and "costs" were therefore taken as legitimate and resulted in policy changes. Regarding the issue of "health", the Ontario government also showed acceptance towards the movement by addressing this concern in 2009 by funding research on the health effects of renewable energy and wind turbines and releasing an official report on the issue (CMOH 2010).

Though the Ontario government has not adopted all the demands of the anti-wind movement, its response can be characterized as a major policy shift as it included changes to the support system (from a FIT to an auction system) and the decision-making system (including better municipal involvement). It can therefore be classified as an *integrative* response.

The power of anti-wind movements: explaining government responses with the concept of discursive energy space

Despite relatively similar movement characteristics, the anti-wind movements in Brandenburg and Ontario faced different government responses. This study asked for the role of context in disputes over wind turbines and conceptualized the discursive energy space as the discursive and the institutional background of such a dispute. How did the meaning context and the formal institutional context thus influence the dispute over wind turbines in Brandenburg and Ontario? Which power effects can we identify?

The meaning context and movement discursive strength

The meaning context of the discursive energy space was conceptualized as encompassing discourses on energy transitions and more general framings of the energy sector. While such a broad conceptualization may include a myriad of different discourses and framings, the emphasis is here on the discourses and framings that stood out to have an impact on each movement's discursive strength. Those are the *Energiewende* and *Energieland* discourse in Brandenburg and the health discourse in Ontario. They had an impact on how the government addressed movement concerns, on the support they received from the opposition party and on mobilizing activities.

The discourses of Energiewende and Energieland weakened Brandenburg's anti-wind movement

Brandenburg, as a German federal state, is strongly tied to the national *Energiewende* discourse which, until the 2017 amendment of the EEG, implied that the energy transition discourse lies at the intersection between *ecological modernization* and *democratic pragmatism*. The *Energiewende* discourse originated in a strong anti-nuclear social movement and involves a nuclear phase-out in the short term and coal phase-out in the mid-term, the development and integration of renewable energies, ambitious goals on greenhouse gas reduction, and the improvement of energy efficiency (Hake et al. 2015). Solar, wind and biomass are therefore not only to replace nuclear, but in the light of climate change they are also anticipated to ultimately replace Germany's coal fleet. The *Energiewende* discourse is strongly anchored in German society and is strongly positively connoted. It makes *fossil discourses*, which often openly dismiss the science of climate change, a marginalized position, although their importance has recently been growing.[2]

The existence of the *Energiewende* discourse negatively impacted the ability of the Brandenburg anti-wind movement to have the state government address their concerns and accept them as legitimate spokespersons. This made it more difficult for the Brandenburg movement to scale-up their concerns. The then ruling coalition of SPD and DIE LINKE had embarked on an *ecological modernization* strategy to considerably increase the rate of renewables in Brandenburg's energy system and ultimately replace its coal sector. As the ultimate goal of Brandenburg's state government was to build an industry comparable to its coal sector, halting this development by adopting the anti-wind movement's demands would not have been in line with the goals of the *Energiewende* to phase out nuclear and eventually coal throughout Germany. The commitment to the *Energiewende* is further strengthened in Brandenburg by the discourse on *Energieland*. In its Energy Strategies 2020 and 2030, the government framed Brandenburg as *Energieland* and "energy export state" (MW 2008; MWE 2012). This refers to the historic importance of Brandenburg's coal sector as a major energy producing industry during GDR times and emphasizes the ambition to sustain an economy based on energy generation even beyond Brandenburg's future coal phase-out. The framing of *Energieland* with its focus on energy export was also endorsed in the government's reaction to the anti-wind protest and was used to reject their demands and tackle the criticism linked to the question of why Brandenburg needs to have more wind turbines than other federal states (MWE 2012, Interview Ralf Christoffers). Implementing larger setbacks, and thus practically reducing further wind turbine development was not regarded as in line with Brandenburg's "being" and "remaining" an *Energieland*. Brandenburg's framing as *Energieland* weakened the anti-wind movement's ability to have the state government adopt its concerns.

The *Energiewende* discourse in Brandenburg also mattered for the support the anti-wind movement enjoyed from the major opposition party, the CDU. The CDU parliamentary group engaged in a number of parliamentary initiatives to support the demands of the popular initiative and urge the state government to make use of the temporary change in the Federal Building Code (*Länderöffnungsklausel)* to introduce larger standardized setbacks. Nonetheless, the CDU group later deviated from the "10h" position and asked for a setback of 1,500 meters while continuing to endorse wind turbines as a sustainable technology in general. Furthermore, the CDU group did not support the anti-wind movement to a similar degree as the Progressive Conservatives in Ontario. The CDU, for example, did not make wind turbines a central topic in the run-up to state elections. The reason for this limited support can again be found in the importance of the *Energiewende* discourse in Brandenburg. As the CDU spokesperson for energy in the state parliament reported, he had the impression that many anti-wind groups follow their private interest and do not consider the public interest. Politics, on the other hand, must balance between different needs of the public good and not follow individual interests (Interview, Dierk Homeyer). Furthermore, he described many of the wind opponents he met as outright climate skeptics, bringing along folders full of "evidence" to the meetings that allegedly showed the non-existence of climate change. While the interviewed local wind opponents in the Temnitz region did not oppose the *Energiewende*, the umbrella organization *Rettet Brandenburg* is a member of the German nationwide umbrella organization *Vernunftkraft*, which openly dismisses the ideas of the *Energiewende* and climate science. At one of their rallies in Potsdam, *Rettet Brandenburg* invited a speaker of *Vernunftkraft* who openly put forward climate skeptic theories (Participant observation in Potsdam). The importance of the *Energiewende* discourse in Brandenburg's meaning context thus contributed to a weaker discursive strength of those following *fossil discourses*, which disqualified parts of the anti-wind movement as legitimate spokespersons from the CDU's point of view.

Moreover, the importance of the *Energiewende* had impacts on mobilizing activities of the local anti-wind group. Although the interviewed grassroots organization was generally in favor of wind turbines and the *Energiewende* and only objected to the vast dimension their development has taken in Brandenburg, they reported that it was difficult to collect petition signatures and gain local support for their cause. For instance, when one activist approached a neighbor to ask for support against wind turbines in the area, the neighbor asked whether she preferred having a nuclear reactor in the neighborhood instead. This anecdote and others suggest that the prevalence of the *Energiewende* discourse effectively diminished the discursive strength of Brandenburg's anti-wind movement. It weakened their arguments and their recognition as legitimate spokespersons towards the government, the major opposition party and also concerning local mobilization.

The discourses of *Energiewende* and *Energieland* thus weakened Brandenburg's anti-wind movement, while the absence of a similarly strong energy transition discourse strengthened Ontario's movement. Ontario's GEA was introduced following the discourses on *ecological modernization* in light of Ontario's coal phase-out and the 2008 economic crisis and *democratic pragmatism*, as the GEA was also intended to spur small-scale energy generation. The supply gap projected to be left by Ontario's coal phase-out constituted an important window of opportunity for the GEA Alliance to call for the policy. There was however no intention to phase out Ontario's nuclear sector. The aim behind Ontario's renewable energy program was thus rather capacity-driven than technology-driven as compared to Germany and Brandenburg. Consequently, it was easier for the Ontario government to address the anti-wind movement's concerns and crucially curb the renewable energy program, since it was only intended to provide additional capacity for an energy system based on mainly nuclear and to a lesser degree hydro-electric power.

In Ontario, the meaning context on energy transitions is therefore inherently different. While Ontario's 2014 coal phase-out was informed by the discourse on ending air pollution and was supported by society (Harris et al. 2015), ending nuclear power is not part of this energy transition discourse. By contrast, nuclear power enjoys high levels of public support and the refurbishment of nuclear reactors is a major strategy in Ontario's Long Term Energy Plans. Ontario's 2013 Long Term Energy Plan (LTEP) called "Achieving Balance" anticipated one half of Ontario's energy mix to be made up of nuclear, while the other half should be generated by renewable power including hydro-electric (Ontario Ministry of Energy 2013). Wind power is thus regarded as an additional source of power. The 2013 LTEP thus aimed at "achieving balance" by fostering a 50 percent share of each nuclear power and renewables, but no further role of renewables was anticipated. The 2017 LTEP endorsed this position by making the costs of electricity bills a top priority for Ontario's energy sector and framing nuclear power as one of the most cost-effective solutions (Ontario Ministry of Energy 2017). The plan further announces that long-term contracts such as the feed-in tariff for renewables would not be pursued. While renewables are thus regarded as an expensive way to procure electricity, refurbishing nuclear reactors is seen less critical.[3] A contributing factor is that nuclear power enjoys broad support and has not been as controversial as wind power in Ontario. This different role of nuclear power in Ontario's meaning context as compared to Brandenburg contributed to the considerably higher discursive strength of the anti-wind movement in Ontario. As the government had the option to endorse nuclear power, the government could easily respond to anti-wind protest by ending policy support schemes for renewable energy. Therefore, choosing refurbishment of nuclear reactors over further wind energy procurement targets can indeed be considered "cheaper" for Ontario politics.

This different discursive background also removed possible barriers to collaboration between the major opposition party and the anti-wind movement. The leader of the Progressive Conservatives tightly worked together with the leader of WCO and ran on a platform of placing a moratorium on wind turbines. He substantially contributed to making wind turbines a central topic of debate in the run-up to the 2011 provincial elections. He fully supported the anti-wind movement and even called his party "the voice for the anti-wind side" (Interview, Tim Hudak).

The absence of a similarly strong discourse on energy transition to the German *Energiewende* discourse had another consequence for the local level. Anti-wind protest in the local case studies of West Lincoln and Wainfleet can be depicted as an "information battle" with anti-wind activists identifying as "information crusaders" because wind turbines were not linked to a positively connoted discourse such as *Energiewende*. Instead, local activists complained to not have received the requisite information about the wind development, its effects and reasons for its development (Focus group discussion, Ontario). Wind turbines were framed as a "risk technology" and information from the government and developers was not trusted. The group expressed the view of being the ones with the "true" information and being denied the correct details on the projects. This viewpoint of "wrong" versus "correct" information was much more forcefully expressed in Ontario than in Brandenburg, where local anti-wind activists strongly identified the *Energiewende*, next to a profit-seeking wind industry, as the central driver behind wind turbine development. While the existence of the *Energiewende* discourse thus contributed to a general understanding of why wind turbine development was supported by the government, a similar understanding was missing in the Ontario local case study.

This had effects on the mobilizing activities of the anti-wind movement, which is also important for the movement's discursive strength, as it is more likely that the more people join a movement, the more likely their demands are to be met by the government. As discussed earlier, the existence of the *Energiewende* discourse made it more difficult for the Brandenburg group to mobilize others to join their group. The fact that the anti-wind movement in Brandenburg had difficulties in garnering support is also evident in their missed target of mobilizing enough people to register their vote for the public referendum in 2016, calling for a "10h" setback and no wind turbines in forests. Due to the absence of a similarly strong discourse in Ontario, the group should have been more successful in mobilizing others. Yet, the group also reported that it was difficult to mobilize people in general, which they attributed to peoples' general laziness and the prevailing overall information deficit on wind turbines (Focus group discussion, Ontario). It can still be assumed that mobilization and support for the anti-wind movement in general benefitted from the absence of a positively connoted discourse supporting wind developments. This is for example visible with the "unwilling host movement", which encompassed Ontario municipalities that put forward motions declaring themselves unwilling host to wind turbines (Unwilling Hosts, 2015).

Ontario's meaning context strengthened the "health" frame of its anti-wind movement

Another aspect of Ontario's meaning context crucially informed the anti-wind movement's struggle and gave a sense of urgency to their framing of being denied their rights. Adverse health impacts are prominently represented in Ontario's meaning context and allowed the anti-wind movement to successfully frame wind turbines as a potential health threat.[4] The assertion that a wind turbine installation possibly or actually has severe negative impacts on the well-being and overall health of people living in its vicinity is a very strong argument. It raises worries and fears, depicting wind turbines as a risky technology. This framing further undermines the green image of wind turbines. This argument is also in line with the common storyline that the rural population is disadvantaged to the benefit of the urban population. Consequently, if this argument is considered as legitimate and enters the debate at the provincial or state level, this suggests that the anti-wind movement has made a considerable step towards putting pressure on the government to address its concerns. Furthermore, the successful depiction of wind turbines as a health threat enhances the mobilization potential of anti-wind groups.

As the local case study in West Lincoln and Wainfleet in Ontario showed, members of the local grassroots anti-wind organization expressed deep concerns about their well-being and felt helpless about becoming exposed to wind turbines (Focus group discussion, Ontario). Anti-wind activists also referred to the 2009 book on the "wind turbine syndrome" by a US-based nurse (Pierpont 2009) (Focus group discussion, Ontario; Interview, John Laforet). This publication associated wind turbines with a number of health issues, such as nausea or sleep disturbance, and became a major point of reference in the Ontario debate. The discussion on the wind turbine syndrome was reflected in anti-wind websites from different countries and spread over the internet. As the local case study revealed, the internet was a major source of information and for some of the interviewees, it contributed considerably to the forming of their opinion. One of the interviewees referred to Pierpont's book and reported to have experienced the described syndromes. Hence, anti-wind activists in Ontario were confronted with readily-available online information from other groups that possibly exacerbated their concerns.

While the health concern was also important in the Brandenburg case study in the Temnitz region (Focus group discussion, Brandenburg), the issue of adverse health impacts only became a matter of provincial concern in Ontario, but not in Brandenburg. As a result, the Ontario government funded research on wind turbines and released an official statement on health impacts and wind turbines in 2010 (CMOH 2010). The leader of the Progressive Conservatives called for a moratorium on wind turbines until the Health Canada study, a further study on the health aspects of wind turbines announced by the federal Ministry of Health in 2012 (Government of Canada 2014) was completed. The two case study municipalities of Wainfleet and West Lincoln also adopted motions to call upon the government to install such a moratorium.

How did the Ontario meaning context contribute to this successful framing of wind turbines as a health threat, in contrast to Brandenburg where health concerns were also raised by opponents but the argument was not successfully scaled-up to the provincial level? This can be explained by the different importance of the health frame in the meaning contexts of the two jurisdictions. In contrast to Brandenburg, health issues were an important component of Ontario's meaning context. This is exemplified by the fact that media articles consistently linked wind turbines to adverse health impacts. Reporting intensified around the year 2009 when the GEA was enacted (Deignan et al. 2013; Hill and Knott 2010; Songsore and Buzzelli 2014). Whether media reporting followed the general public perception or actually primarily contributed to the formation of opinion and reinforced existing worries cannot easily be determined. Most certainly, both are true. Yet, the importance of the issue in media coverage shows that the "adverse health impacts frame" has high public salience in Ontario's meaning context. Further evidence is that the Environmental Review Tribunal, the official process in which opponents may appeal a specific project, acknowledges the issue of "human health" next to "environmental damage" as the only legitimate reasons to appeal a project. The importance of the health frame is thus not an exclusive product of media coverage. Indeed, the health frame in the realm of the energy sector has historically been a crucial component in Ontario's meaning context.

Adverse health impacts had already been the major framing of the political campaign for Ontario's coal phase-out. The *fossil discourse* in terms of a largely unquestioned use of coal-fired electricity began to weaken around the year 2000, when the grassroots-led campaign to phase out coal had been started and was positively received by the public. The *fossil discourse* was openly challenged by a discourse that saw coal combustion as the major reason for air quality concerns and corresponding adverse health impacts. This contributed to the framing of energy production in terms of health, producing a storyline that regarded energy production, especially coal, as a potential health hazard. Responding to this public storyline of energy production as a public health issue, the Progressive Conservatives proposed the first procurement targets for wind power, solar and biomass in Ontario in the early 2000s. The link between energy generation and health issues was therefore made early on and has persisted since. Correspondingly, the introduction of renewable energy policy by the GEA was also framed as a measure to fill the supply gap of the coal phase-out with a "healthy" source of power.

In Brandenburg, local wind opponents raised the issue of health to a similar degree as in Ontario. Both groups complained about not knowing where to direct their concerns and not being taken seriously by the local authorities. While the Brandenburg interviewees compared wind turbine noise to "torture", Ontario interviewees were convinced that they are victims of human experimentation (Focus group discussions, Brandenburg and Ontario). Despite the similar force of health concerns at the local level, health issues were only implicitly included in the Brandenburg umbrella organization's call for a setback of "10h", but there was no concerted call for a moratorium until health studies were completed as in Ontario. Furthermore, health concerns did not figure as prominently in the debate over wind turbines at the Brandenburg state level as in Ontario. One could argue that the health discussion in Brandenburg was less important because regional planning usually determines the setback requirements for wind turbines at approximately double the distance of the Ontario minimum setback of 550 meters. The shorter setback could have intensified the debate over health impacts in Ontario, because shorter setbacks have more effects on those living close to wind turbines than larger ones. Yet, this would not serve as an explanation why anti-wind opposition was much less common in the beginnings of wind turbine development in Brandenburg when setbacks were much shorter than the current 1,000 meters. Furthermore, this does not explain why both local anti-wind groups were similarly concerned over health issues. A better explanation of why health concerns were not successfully scaled-up to the state policy-making level as in Ontario can be found in the fact that health issues did not constitute a central component of Brandenburg's meaning context.

The issue of health was not linked to the energy sector as in Ontario and discussions about "wind turbine syndrome" did not gain much ground in Brandenburg. This may have been because of a language barrier and suggests the stronger strength of the health argument in the North American context, where internet coverage in English certainly contributed to the salience of this argument. In conclusion, the Brandenburg movement did not successfully scale-up health concerns to the state level like the Ontario movement did. As the health argument adds to discursive strength, the Ontario movement therefore generally benefitted from the importance of the health argument in Ontario's meaning context.

The power effects of the meaning context

Overall, the power effects of the meaning context provided more discursive strength to the anti-wind movement in Ontario than to the movement in Brandenburg. The following specific aspects of power can be identified. The different meaning contexts in Brandenburg and Ontario contained different discourses, which attributed more or less weight to the movements' framings and influenced the way the government addressed their concerns, the support they received from the political opposition party, and their chances of mobilization. This power effect of the discourses of the meaning contexts relates to Foucault's notion of discursive power, which posits that each society has some "types of discourse which it accepts and makes function as true" (2000, p. 131). Foucault originally referred to more robust and long-term societal discourses, while the meaning context here also includes more volatile discourses that can change in the medium-term. Despite his focus on longer-term societal macrostructures, his conception of discourse power has also informed the study of medium-term energy and sustainability transitions (Bues and Gailing 2016; Partzsch 2015) and has been used in social movement studies (Baumgarten and Ullrich 2016). The *Energiewende* discourse in Brandenburg represented such a powerful discourse that merged with Brandenburg's depiction of itself as an *Energieland*. In Ontario, the framing of wind turbines in terms of adverse health impacts was successfully raised by the anti-wind movement because this framing also represented a powerful and generally accepted discourse in society. These powerful discourses either provided strength to the anti-wind movement to raise the topic (Ontario) or to the government to reject the movement's demands (Brandenburg).

The second face of power is also relevant here. A government confronted by a movement may directly refer to the prevailing discourses and framings in the meaning context to reject a movement's demands. In Brandenburg, politicians justified their rejection of the demands because of the perceived need for Brandenburg to remain an *Energieland*. This means that the discussion of larger setbacks and banning wind turbines in forests was foreclosed by prioritizing the presumed larger goal of Brandenburg's electricity production. This can be termed as an instance of the second face of power, which is used to "limit the scope of the political process to public consideration of only those issues which are comparatively innocuous to A" (Bachrach and Baratz 1962, p. 948). This form of power to limit the scope of what can be discussed has also been termed as the context-shaping form of power (Hay 1997).

In conclusion, the power effects of the meaning contexts in Ontario and Brandenburg had different outcomes on the movements' discursive strength. While the meaning context in Ontario contributed to the anti-wind movement being able to successfully scale-up their concerns, the Brandenburg movement was less able to do so. The different shape of the meaning context therefore contributed to considerably more pressure on the Ontario government than in Brandenburg. The argument can be summarized with the following conclusion: The more a movement's framings correlate with prevailing discourses, the easier it is for the movement to scale-up its demands.

The formal institutional context and movement discursive strength

Anti-wind movements do not only gain discursive strength from the prevailing meaning context, but also from the formal institutional context. The framework of discursive energy space introduced the formal institutional system as comprising the role of the federal energy policy and the local space of participation. While the former depicts the way in which the sub-national wind energy policy is embedded in the overall federal energy policy framework, the latter refers to how the local planning system for wind turbine siting is devised and implemented.

When faced with anti-wind movements, a government's intention may not primarily be to respond in exclusionary terms by denying the legitimacy of the movement and dismissing their concerns. Governments may rather be open for feedback in order to accordingly improve their policies or to at least appear responsive to their constituents. In both Brandenburg and Ontario, the sub-national governments have indeed engaged in dialogue with the population, for instance in the context of the revisions of the FIT and the Long Term Energy Plan (Ontario) and in the form of the "energy tours" of the then Minister of Energy (Brandenburg). Yet, governments may not want to yield to all demands of anti-wind movements because this may compromise their policy targets that are possibly supported by a larger part of the constituency than the social movement represents. At the same time, acknowledging a social movement as legitimate while declining to meet its demands is a difficult undertaking. The degree to which the contested sub-national policy is embedded in federal energy policy provides opportunities for this *diplomatic* government response, as decision-makers being confronted with anti-wind challengers may refer to their sub-national jurisdiction's responsibility towards the federal level. This opportunity therefore weakens the ability of anti-wind movements to scale-up their concerns to the sub-national level where the contested wind policy might be changed. This power effect of the formal institutional system provided more discursive strength to the anti-wind movement in Ontario than in Brandenburg.

Brandenburg's link to federal energy policy weakened movement discursive strength

Brandenburg's renewable energy sector is strongly shaped by the German federal level. The FIT scheme of the federal Renewable Energy Sources Act (EEG) provides the overall financial support scheme upon which wind energy has flourished in Brandenburg. Furthermore, Brandenburg must deliver on national and European targets for renewable energy procurement and greenhouse gas emission reduction. Already in 1994, Brandenburg referred to the national CO_2 reduction targets to motivate support for wind energy development (MUNR 1994). Notwithstanding this important national impetus, Brandenburg has pro-actively steered and supported wind energy development via its planning approach and via a number of research and development programs to a larger extent than other federal states. The link of Brandenburg's energy policy to federal energy policy provided the government with opportunities for a *diplomatic* government response and therefore weakened the Brandenburg movement's discursive strength.

Indeed, when the Brandenburg government was confronted with anti-wind opposition, it straightforwardly acknowledged those concerns that could only be solved at the national level. For instance, the government endorsed the changes made to the EEG at the federal level, which comprised a change from the FIT to an auction system and a limitation of further wind energy procurement. Support for this reform was motivated by the expectation that it would tackle rising electricity costs, one of the issues the Brandenburg government considered a major source of public discontent. The reduced rate in new wind energy development resulting from the amended EEG was expected to provide time to catch up with developing the necessary electricity grids and hence decrease costs for Brandenburg residents. The Brandenburg government also called upon the federal level to change the rule imposing the costs for local grid development upon the population of the area where this grid development occurs, which severely disadvantages rural areas with many wind turbines. Furthermore, responding to anti-wind protest comprised an information strategy which involved the reference to Brandenburg's obligations to contribute to the federal renewable energy targets and the *Energiewende* in general (Interviews, Ralf Christoffers and ministry representative). This also reflects the importance of the *Energiewende* and *Energieland* discourse, which were discussed earlier under the meaning context, and underscores the fact that the formal institutional system and the meaning context are interrelated. The Brandenburg government thus commented on or interacted with the federal policy level in several ways. It thereby conveyed the impression that it considered anti-wind sentiments stemming from higher electricity costs as legitimate and also showed that it acted on that issue, even if the local case study suggested that costs were not actually among the major points of concern (Focus group discussion, Brandenburg).

In conclusion, the embeddedness of Brandenburg's energy system with federal energy policy renders *diplomatic* government responses to anti-wind protest more feasible. This is because the government can easily agree with the social movement's demands and still legitimate its inaction in adopting them by referring to either the federal level's responsibility for tackling the concerns or by referring to the necessity to deliver on federal renewable energy targets. In fact, by endorsing the changes to the EEG from a FIT to an auction system at the federal level, the Brandenburg government could easily put forward a *market rationalism* discourse without having to change the policy support system themselves. A *market rationalism* discourse regards market competition as the major mechanism to achieve environmental protection and endorses competitive procurement schemes for renewable energy. The Brandenburg government thus pro-actively used the formal institutional system to endorse its wind energy program in the face of protest. This was different in Ontario.

Ontario's sovereignty over energy strengthened the anti-wind movement

In contrast to Brandenburg, the contested wind energy policy in Ontario, the 2009 GEA, was not part of a larger federal strategy for emissions reduction or renewable energy deployment. Although Ontario has sovereignty over its energy sector, the federal level could still have served as a motivating actor in the context of a Canadian strategy to curb emissions. The 2016 Pan-Canadian Framework on Clean Growth and Climate Change (Government of Canada 2016) presented under Premier Justin Trudeau is an example of such a framework, building upon the commitments and actions of the Canadian provincial governments. However, the Canadian federal government under Stephen Harper, which was in power during the time period under study, had a reluctant stance on energy system change towards decarbonization and renewable energy. The federal government's inaction on climate change mitigation culminated in the December 2011 announcement of Canada's official withdrawal from the Kyoto protocol, the preceding UNFCCC agreement to the 2015 Paris Agreement. This illustrates that Ontario's coal phase-out initially planned for 2007 and eventually realized in 2014, as well as the 2009 GEA, fell in a period of weak federal leadership in climate protection. The fact that Ontario has acted independently from the federal level has important consequences for the discursive strength of Ontario's anti wind-movement. Faced with challengers, the Ontario government could not refer to the federal level to defend its wind energy policy in the face of protest like in Brandenburg. This strengthened the anti-wind movement because the Ontario government could be depicted as the only ones responsible for the unwanted wind turbine development and the government could not defer responsibility and use a *diplomatic* government response.

The fact that Ontario's renewable energy policy was not part of strong federal leadership on renewable energy policy also meant that it was more vulnerable to party politics. Ontario's GEA was endorsed by and associated with the Liberals, who had most of their electoral support in Ontario's urban areas. By contrast, the Progressive Conservatives, as the major opposition party and proponents of the *market rationalism* discourse, disregarded the idea of supportive renewable energy legislation and had most of their electoral support in the rural areas. Due to this electoral divide between the urban and rural areas, the GEA was also depicted as an "urban Liberal" project against the rural Progressive Conservative voters (Walker et al. 2018). Given that no overarching federal legislation backed the GEA made the policy more vulnerable to being reversed in the event that the opposition party won an election. This is further aggravated by the fact that in majoritarian electoral systems such as Canada, there are fewer parties in parliament and politics, including debates about energy transitions, is often more confrontational (Lockwood et al. 2016). This certainly added a different note to the disputes over wind turbines in Ontario than seen in Brandenburg's proportional representation electoral system.

The pressure to act on the opposition's demands that were seemingly backed by a forceful anti-wind movement consequently put more pressure on the Ontario government. By suspending the FIT scheme for large-scale projects and introducing a system that better included the municipality's perspective in 2013, the Ontario government in fact responded in an *integrative* way by incorporating the two major demands of the Progressive Conservatives, namely the concerns on costs and poor municipal involvement. By doing so, they effectively took wind out of the opposition's sails to potentially overturn the GEA in case they won the next election.

In conclusion, the way in which the respective wind energy policy was embedded in federal legislation strengthened the anti-wind movement in Ontario more than in Brandenburg where the government could actively pass some demands onto the federal level while rejecting others. This suggests that the embeddedness of a sub-national policy in the federal context can provide some stability in the face of differing discursive preferences at the sub-national level. The discussion also suggests that the absence of such a framework in favor of renewable energy and a corresponding meaning context may also reinforce competition between the ruling and the opposition party, making it more likely that the ruling party yields to the opposition's demands. The question arises of whether Brandenburg would have responded in the same way to the anti-wind movement as Ontario if it had the same legislative authority and was not subject to an overarching legislative framework. Since the Brandenburg government officially endorsed the changes to the federal EEG, which included a shift from a FIT to an auction system ("EEG 2017"), the government would probably also have shifted their FIT system. Still, the government would most likely not have lowered the new procurement targets for wind energy like Ontario did. This is because the *Energiewende* and *Energieland* discourse were too strong to curb the renewable energy program to the same degree as in Ontario. This suggests that the meaning context and the formal institutional system are interrelated. This will be discussed in further detail later. Before the precise power effects of the different embeddedness in federal energy policy are presented, the following section first introduces the effects of the space of participation on movement discursive strength.

The space of participation strengthened the movement in Ontario ...

The space of participation impacts the degree to which the anti-wind movement can scale-up their concerns to the policy-making level in two major ways. First, the more restrictive the space of participation is perceived as by the anti-wind movement, the more forcefully it may react. If the characteristics of the space of participation allows the movement to invoke certain frames that mobilize support, this may increase a movement's discursive strength. Similarly to the previous discussion on the role of federal energy policy, the space of participation may also provide opportunities for a government confronted with anti-wind challengers to respond in a *diplomatic* way instead of dismissing the movement's claims altogether. Government politicians may accordingly refer to the space of participation to address particular concerns, delegating their response. This weakens the ability of anti-wind movements to have their demands addressed at the sub-national government level that has the capability to change the overall policy or the space of participation as demanded by the anti-wind movement. How did the space of participation thus impact movement discursive strength to scale-up its concerns in Brandenburg and Ontario? As the effects of Ontario's space of participation on movement discursive strength were more obvious than in Brandenburg, Ontario will be discussed first.

The space of participation in Ontario encompassed open house meetings organized by the private wind turbine developer, the opportunity for the residents to submit comments to the ministry before operating approval to the proponent is granted, and the appeal hearings of the Environmental Review Tribunal. Superseding the municipal level, this streamlined approach replaced the existing planning system for wind turbine development in which municipalities had a stronger role. The new system was introduced *upon* implementation of the GEA. Although the change of the implementation system was due to a variety of reasons, only one reason dominated the public debate. The then Prime Minister Dalton McGuinty introduced the act by his much-cited remark that "NIMBYism will no longer prevail" (Ferguson and Ferenc 2009). This statement contributed, next to the actual change of superseding the municipalities, to a framing of wind turbine development as "top-down", "undemocratic" and even "dictatorial" (Interviews with township councilors and focus group discussion Ontario). This had important consequences on the nature of the emerging local protest.

The changes to the space of participation, including its negative framing, caused intense opposition (Hill and Knott 2010; Walker and Baxter 2017). It mobilized strong protest from the municipal level exemplified by numerous township councils officially declaring themselves as "unwilling host to wind turbines" (Unwilling Hosts 2015). As the case study of West Lincoln and Wainfleet in Ontario showed, both municipalities were strongly opposed to being superseded in wind turbine decision-making. The opposed councilors closely cooperated with the anti-wind groups and engaged in municipal action to display their opposition (Interviews, with township councilors). This further supported and strengthened the anti-wind movement. Another exacerbating effect is that the Environmental Review Tribunal, as the last possible step of stopping an unwanted project, only deemed two arguments as legitimate. These were impacts on human health and irreversible damage to the environment. The fact that no further aspects were acknowledged and only costly expert evidence was accepted to prove health or environmental damage further contributed to the opponents' framing of the space of participation as undemocratic (Focus group discussion,

Ontario). The corresponding storyline referring to "democracy" and "rights" was invoked by the anti-wind umbrella organization to lever wind opposition to the provincial level, where the major opposition party responded to it positively. The argument therefore had a powerful effect on garnering support at the policy-making level.

... and weakened protest in Brandenburg

In contrast to Ontario, Brandenburg's space of participation principally has the characteristics of a locally informed inclusive process. Wind power planning is carried out at the regional planning level in which five regional planning authorities designate wind suitability areas following a set of "soft" and "hard" criteria. The process generally allows for locally adapted solutions to address particular raised concerns. In the case study of Prignitz-Oberhavel, the regional assembly for instance included the provision that wind turbines must not encircle a village by more than 180 degrees. In Ontario, by contrast, there was much less opportunity for the local or regional level to pro-actively determine specific rules and conditions that wind energy developers needed to abide by.

Notwithstanding this example of local involvement in Brandenburg, numerous legal changes at the European and national level and corresponding court rules have considerably restricted the autonomy of the regional planning authorities (RPGs). Examples include the 1997 change in the Federal Building Code that classified wind turbines as "privileged facility", the 2008 European Directive on mandatory public participation or the 2010 rule of the Higher Administrative Court Berlin-Brandenburg that specified the identification of wind suitability areas and imposed a strict procedure upon regional planning. As the head of the public planning office in the local case study stated, better compromises between municipal and regional planning were achieved before the 2010 court rule (Interview, Ansgar Kuschel). Next to these legal changes from the European and national level and subsequent court rules, ministries in Brandenburg have advised the RPGs to support wind turbine planning by issuing several directives. The instructions did not restrict participation in the form of excluding particular levels from decision-making such as the municipal level in Ontario, but prescribed specific area targets for designating wind suitability areas (e.g. MIR and MLUV 2009) or streamlined the procedure by lowering the required separation distances to habitats of specific species (e.g. MUGV 2011). The European, national and state requirements that regional planning faces has severely limited the actual ability to find local compromises and has substantially constrained its room of maneuver (Interview, Ansgar Kuschel).

The fact that local participation opportunities in the regional planning process in Brandenburg are heavily restricted would in principle serve as a major mobilizing argument for the anti-wind movement. The movement could have focused on the argument that regional planning is an overly complex system that only partially responds to local and regional planning input, but is instead informed by European and national legislation and state guidance. While the local anti-wind group in Temnitz raised such concerns (Focus group discussion, Brandenburg), this aspect was not scaled-up as a major issue to the sub-national level. Comparing Brandenburg's space of participation to Ontario's offers possible explanations.

In contrast to Ontario, the local level in Brandenburg has not been directly superseded by one single legislative action like the GEA in Ontario. To a limited degree, local input is possible in Brandenburg and the changes that made the space of participation more restrictive for local input occurred along a longer time period and did not directly originate from the sub-national government. Furthermore, despite the numerous changes stemming from different political levels which considerably restricted the autonomy of the RPGs to find locally adapted solutions, Brandenburg's space of participation has continuously been framed as being a locally decided and inclusive process. As early as 1996, the government framed the regional planning process as a win-win situation between nature conservation and environmentally benign renewable energy development (MUNR 1996). This positive framing has since persisted and is reflected in various governmental policy documents (e.g. SPD Brandenburg 2015; SPD Brandenburg and DIE LINKE Brandenburg 2014) and also played a role in the resolution the state parliament adopted in response to the demands of the popular initiative *Rettet Brandenburg* (Landtag Brandenburg 2015).

This ongoing positive framing of the space of participation and its continuous changes towards a more restrictive process over the course of time had several important effects on the strength of the anti-wind movement. First of all, the depiction of the space of participation as "unjust" did not reach similar force in Brandenburg as in Ontario because the changes made to it were less visible and did not occur all at once. Although the anti-wind opponents in the local case study regarded the regional planning process as difficult to understand and unfair with regard to the need to provide costly expert evidence (Focus group discussion, Brandenburg), *procedural justice* arguments were not scaled-up to the provincial policy-making level. The space of participation in Brandenburg therefore only gave rise to the powerful injustice argument to a limited degree. Second, the positive framing of Brandenburg's space of participation as a local and inclusive process allowed the government to endorse the existing system of regional planning when they were confronted by anti-wind challengers. Government representatives could refer to the system that was already in place and did not have to legitimate their choice of installing it, as was the case in Ontario. This allowed for a *diplomatic* response from the Brandenburg government.

The power effects of the formal institutional context

Given the different effects that federal energy policy and the space of participation had on the anti-wind movements in Brandenburg and Ontario, how can the power effects of the formal institutional system be conceptualized? The space of participation had indirect effects on the chances of anti-wind movements to succeed by potentially providing them with powerful arguments that elicit support. This is connected to the first and second face of power. The first face of power in which A can "get B to do something that B would not do otherwise" (Dahl 1957, p. 201) refers to the fact that the space of participation may restrict peoples' involvement directly by determining who is eligible at which stage to have a say in the process. The more directly and obviously the first face of power as constraining people's input is played out via the space of participation, the heavier the reaction of the anti-wind movement may be. This was the case in Ontario, where the space of participation directly excluded the municipal level from decision-making over wind turbines. This change sparked a strong discourse focusing on "rights" that also mobilized support from the opposition party and from the municipal level, in form of the unwilling-host-movement. The characteristics of the space of participation in Ontario therefore supported the anti-wind movement to invoke the injustice frame, which added to the movement's discursive strength and consequently increased pressure on the government.

The second face of power mattered both for the impacts of federal energy policy as well as the effects of the spaces of participation. This form of power also involves the conscious influencing of values or political institutions, but the analysis here focuses on the aspect of non-decision-making power in discursive terms. By referring responsibility to another level of decision-making, another level of government or to non-governmental agencies, this form of "non-decision power" restricts the scope of governmental decision-making and has also been called "strategic institutional design" or "institutional" or "governmental depoliticization" (Flinders and Buller 2006; Flinders and Wood 2014; Hay 2007; Landwehr and Böhm 2014). This does not necessarily mean that those engaging in this form of institutional depoliticization do this with conscious intent to restrict decision-making and keep others out. In fact, policy-making, especially in multi-level governance arrangements, is always shaped by decisions being taken at different levels. Yet, employing the concept of "non-decision-making power" and the notion of depoliticization is useful here to illustrate the possible effects of the formal institutional system on a movement's discursive strength.

The fact that Brandenburg's energy policy was linked to the federal level offered opportunities for the government to delegate responsibility to deal with some of the causes of anti-wind protest to the federal level. The Brandenburg government could thus employ "non-decision-making power" to reject the movement's demands. As Ontario has full sovereignty over its energy sector and federal leadership on climate and environmental topics was weak during the studied time period, there were no opportunities for the government to do the same. This non-decision-making power was also relevant for the space of participation and worked in a similar way. As the regional planning process in Brandenburg is positively connoted as a local and inclusive process, the Brandenburg government could easily refer to this level to deal with matters of local anti-wind opposition. This was not possible for the Ontario government. It had framed the space of participation in a negative way as "overcoming NIMBYism" and as the superseding of the municipal level sparked a strong reaction focusing on "rights", the Ontario space of participation was regarded as overly negative by many. It therefore provided little room for *diplomatic* government responses and non-decision-making power, in contrast to Brandenburg.

Concluding this section, the power effects of the formal institutional system in Brandenburg and Ontario had different impacts on the movements' discursive strength. Similarly to the power effects of the meaning context, the formal institutional system strengthened discursive strength of Ontario's anti-wind movement and weakened movement strength in Brandenburg. The argument can be summed up by the following. The formal institutional system may provide opportunities for the government to refer responsibility for dealing with the movement's demands to another level. This makes it more difficult for a movement to scale-up its demands.

The interlinkage of meaning context and formal institutional context

It is important to recognize that the meaning context and the formal institutional system are interrelated and mutually dependent. In fact, the differentiation between meaning context and formal institutional system may be overly schematic, despite being useful for analytical purposes. Several observations can be made from the discussion of the case studies.

In Brandenburg, the main discourses of the meaning context, the *Energiewende* and *Energieland* discourse, were also reflected in the formal institutional context. The German federal energy policy echoed the *Energiewende* discourse including the need for energy system transformation. This need for energy system transformation and the need for the state of Brandenburg to contribute to this goal as reflected in the *Energieland* discourse impacted the way in which the space of participation for wind turbine planning was shaped. The Brandenburg state government actively steered the regional planning process by issuing directives to the regional planning authorities, thereby practically institutionalizing the *Energiewende* and *Energieland* discourse.

In Ontario, a discourse on the need for energy transitions as strong as the German *Energiewende* discourse did not exist, nor was Ontario's energy policy framed or informed by a pro-energy transition federal policy. These aspects could implicitly have contributed to the government's decision to implement a more restrictive space for citizen participation at the local level. If decision-makers had left planning authority to the local communities, this would arguably have halted renewable energy development before it even started due to the absence of an overarching strong discourse for energy transition implying broad public support. Yet, public opposition can also be attributed to the large-scale and often foreign-owned wind turbine projects that were realized instead of community-owned wind projects (Goedkoop and Devine-Wright 2016; MacArthur 2016; Simcock 2016; Wolsink 2007).

The fact that Ontario's contested energy policy was neither embedded in a federal energy policy nor framed by a similarly strong discourse as in Brandenburg had the important consequence that the dispute over wind turbines became a matter of party politics. The Progressive Conservatives supported the anti-wind movement's cause by calling for a moratorium, re-establishing local planning authority and opposing the GEA as a whole. The support of the major opposition party, especially in the run-up to the 2011 elections, considerably contributed to pressure on the Ontario government to react to the movement's demands. In Brandenburg, by contrast, the major opposition party at the time, the CDU, did not become a strategic ally to the anti-wind movement, although they supported some of their concerns. Their lower degree of support was mainly due to the fact that some of the anti-wind activists were depicted as climate skeptics (Interview, Dierk Homeyer), which does not qualify as a legitimate position under the discourse of *Energiewende*. The absence of the support of the major opposition party led to a lower discursive strength of the Brandenburg movement as compared to the Ontario movement.

The power effects of the discursive energy space: summary

Given the described power effects of the discursive energy space, the question arises as to what extent the two sub-national governments could still have acted differently in the face of protest. In other words, is the discursive energy space to be understood as a stiff structure that could have predicted the observed government responses by heavily restricting government autonomy? The answer would be yes if the meaning context represented a rigid fundament staking out the definite discursive playing field of the dispute. Yet, the meaning context on energy transitions is volatile in the medium-term and also influenced by the dispute itself. Ontario's successful coal phase-out for example shows that *fossil discourses* were superseded by discourses in favor of energy transitions. The analysis of the case study in Ontario however suggested that these discourses were not strong enough to weaken the anti-wind movement's discursive strength as was the case in Brandenburg.

This leads us directly to a related question: To which extent do actors have the capacity to "act" or rather "react" in the discursive energy space? To which extent can actors influence the meaning context? Discourses and frames on energy transitions do not exist independently of the actors putting them forward. Actors that form a group and share a set of storylines form a discourse coalition (Hajer 1993; 1995; Mander 2008; Szarka 2004). The investigated anti-wind movements can indeed be regarded as discourse coalitions against wind turbine development in their jurisdiction. The study has shed light on how they form their storylines, how they interact with other actors and how they become active in trying to reach their goals. Yet, each discursive interaction takes place against the backdrop of an existing discursive and an institutional context. A pure perspective on direct discursive interaction and actors alone runs the risk of omitting context. Both prevailing discourses and institutional factors have an effect on the dispute and on how discourse coalitions have power to put forward their goals. Actor-centered approaches such as discourse coalitions are not conflicting to the concept of discursive energy space, but they can be seen as complementary. While discourse coalitions clearly put the emphasis on those putting

Table 6.2 The power dimension of the discursive energy space in Brandenburg and Ontario

Discursive energy space	Relevant face of power	Effect on the anti-wind movement	Empirical findings in Brandenburg	Empirical findings in Ontario
Meaning context	4th face of power	Existence of powerful discourses weakens or reinforces the validity of a movement's concerns.	*Energiewende* and *Energieland* discourse	"Health" discourse, weak overarching discourse on energy transition
	2nd face of power	Those confronted with anti-wind opposition may refer to particular discourses to reject addressing the movement's demands.	(star)	(flower)
Institutional system: link to federal energy policy	2nd face of power	Those confronted with anti-wind opposition may refer to a higher policy level to reject addressing the movement's demands.	Importance of federal energy policy; State government strongly refers to federal level.	Weak federal leadership; No reference to federal level.
			(star)	(flower)
Institutional system: space of participation	1st face of power	The space of participation may directly restrict participation.	Local input is generally possible, but severely limited	Local level is superseded
	2nd face of power	Those confronted with anti-wind opposition may refer to space of participation to reject addressing the movement's demands.	→ No emergence of a strong challenging discourse	→ Emergence of "injustice" and "rights" frame
			(star)	(flower)
	Discursive strength		Moderate →	High →
	Government response		**Diplomatic and exclusionary response**	**Integrative response**

Source: Star: Weakening Movement, Flower: Strengthening Movement.

forward a certain discourse, the concept of discursive energy space investigates how context shapes their capacity to put forward their discourse. The study showed that discourse coalitions can benefit and gain strength from pre-existing discourses and institutional structures as much as they can also shape them.

The concept of discursive energy space can be extended to other issue areas than energy. The concept therefore confirms the notion of political opportunity structure (POS) that deems institutional structures as important for movement outcomes (Bloom 2014; Kitschelt 1986; Vrablikova 2014), as well as generally confirming the notion of discursive opportunities (Aydemir and Vliegenthart 2017; Koopmans and Olzak 2004; Koopmans and Statham 1999). Yet, these two approaches in social movement studies have mostly been studied separately, while the concept of discursive space allows to apply both aspects simultaneously and spells out their power effects. Table 6.2 provides a summary.

The concept of discursive energy space also provides insights for discursive-institutional theory (Schmidt 2008, 2010, 2017). Considering how an actor's capacity to influence others is shaped by different forms of power stemming from given discursive and institutional settings can be insightful for studying how discursive interaction unfolds and how institutions are consequently "constituted, framed, and transformed through discourse" (Campbell and Pedersen 2001, p. 10). This consideration therefore contributes to the core interest of discursive institutionalism. The capacity of actors to have the discursive strength to scale-up their demands and impose their views on others is influenced by the characteristics of the existing discursive-institutional system, which also provides power to those that are targeted. Most discursive institutionalist scholars have referred to the political power of ideas without further specifying how they particularly defined it (Carstensen and Schmidt 2016, p. 318). The concept of discursive space contributes a concept of power to the study of discursive-institutional systems.

Notes

1 The grassroots part of the anti-wind movements investigated here aimed at stopping the proposed project (Niagara Region) or reversing the regional plan which devised wind suitability areas in their area (Temnitz region). Neither of these local groups reached this goal.

2 This is exemplified by the election of the right-wing party, *Alternative für Deutschland* (AfD), which openly dismisses climate science and ran on a platform promising to terminate the German Renewable Energy Sources Act, to the German *Bundestag* in September 2017.

3 However, a cost comparison between Ontario's greenhouse gas reduction options shows that in the coming several years, nuclear power is the most expensive option (CAD 383 per tonne CO_2), followed by solar power (CAD 355), Ontario wind power (CAD 110), Quebec wind power (CAD 31), while energy efficiency and water power imports from Quebec would qualify as the best options with regard to costs (OCAA 2017).

4 It shall be reiterated here that this study refrains from judging whether arguments about the adverse health impacts of wind turbines are valid or not – "successfully framing" means that the government reacted to this framing.

References

Aydemir, Nermin; Vliegenthart, Rens (2017): Public Discourse on Minorities. How Discursive Opportunities Shape Representative Patterns in the Netherlands and the UK. In *Nationalities Papers* 4 (2), pp. 1–15. DOI: 10.1080/00905992.2017.1342077.

Bachrach, Peter; Baratz, Morton S. (1962): Two Faces of Power. In *American Political Science Review* 56 (04), pp. 947–952. DOI: 10.2307/1952796.

Baumgarten, Britta; Ullrich, Peter (2016): Discourse, Power, and Governmentality. Social Movement Research with and beyond Foucault. In Jochen Roose, Hella Dietz (Eds.): Social Theory and Social Movements. Mutual Inspirations, pp. 13–38. Wiesbaden: Springer VS.

Bloom, Jack M. (2014): Political Opportunity Structure, Contentious Social Movements, and State-Based Organizations. The Fight against Solidarity inside the Polish United Workers Party. In *Social Science History* 38 (3–4), pp. 359–388. DOI: 10.1017/ssh.2015.29.

Bues, Andrea; Gailing, Ludger (2016): Energy Transitions and Power: Between Governmentality and Depoliticization. In Ludger Gailing, Timothy Moss (Eds.): Conceptualizing Germany's Energy Transition. London: Palgrave Macmillan, pp. 69–91.

Campbell, John L.; Pedersen, Ove K. (2001): The rise of Neoliberalism and Institutional Analysis. Introduction. In John L. Campbell, Ove K. Pedersen (Eds.): The Rise of Neoliberalism and Institutional Analysis. Princeton, NJ, Chichester: Princeton University Press.

Carstensen, Martin B.; Schmidt, Vivien A. (2016): Power Through, Over and in Ideas. Conceptualizing Ideational Power in Discursive Institutionalism. In *Journal of European Public Policy* 23 (3), pp. 318–337. DOI: 10.1080/13501763.2015.1115534.

CMOH (2010): The Potential Health Impact of Wind Turbines. Chief Medical Officer of Health (CMOH) Report. Available online at www.health.gov.on.ca/en/common/ministry/publications/reports/wind_turbine/wind_turbine.pdf, checked on 10/5/2016.

Cowell, Richard; Bristow, Gill; Munday, Max (2011): Acceptance, Acceptability and Environmental Justice: The Role of Community Benefits in Wind Energy Development. In *Journal of Environmental Planning and Management* 54 (4), pp. 539–557. DOI: 10.1080/09640568.2010.521047.

Dahl, Robert (1957): The Concept of Power. In *Behavioral Science* 2 (3), pp. 201–215.

Deignan, Benjamin; Harvey, Erin; Hoffman-Goetz, Laurie (2013): Fright Factors About Wind Turbines and Health in Ontario Newspapers before and after the Green Energy Act. In *Health, Risk & Society* 15 (3), pp. 234–250. DOI: 10.1080/13698575.2013.776015.

Della Porta, Donatella; Diani, Mario (2006): Social Movements. An Introduction. 2. ed. Malden, MA: Blackwell.

Ferguson, Rob; Ferenc, Leslie (2009): McGuinty Vows to Stop Wind-Farm NIMBYs. In *Toronto Star*, 2/11/2009. Available online at www.thestar.com/news/ontario/2009/02/11/mcguinty_vows_to_stop_windfarm_nimbys.html, checked on 11/2/2015.

Flinders, Matthew; Buller, Jim (2006): Depoliticisation: Principles, Tactics and Tools. In *British Politics* 1 (3), pp. 293–318. DOI: 10.1057/palgrave.bp.4200016.

Flinders, Matthew; Wood, Matt (2014): Depoliticisation, Governance and the State. In *Policy & Politics* 42 (2), pp. 135–149. DOI: 10.1332/030557312X655873.

Foucault, Michel (2000): Truth and Power. In Michel Foucault, Paul Rabinow, James D. Faubion (Eds.): The Essential Works of Michel Foucault, 1954–1984. New York: New Press (The essential works of Michel Foucault, 1954–1984, v. 1).

Goedkoop, Fleur; Devine-Wright, Patrick (2016): Partnership or Placation? The Role of Trust and Justice in the Shared Ownership of Renewable Energy Projects. In *Energy Research & Social Science* 17, pp. 135–146. DOI: 10.1016/j.erss.2016.04.021.

Government of Canada (2014): Wind Turbine Noise and Health Study: Summary of Results. Government of Canada. Available online at www.canada.ca/en/health-canada/services/environmental-workplace-health/noise/wind-turbine-noise/wind-turbine-noise-health-study-summary-results.html, updated on 11/24/2017, checked on 11/24/2017.

Government of Canada (2016): Pan-Canadian Framework on Clean Growth and Climate Change. Canada's plan to address climate change and grow the economy. Gatineau, Québec: Environment and Climate Change Canada. Available online at www.canada.ca/content/dam/themes/environment/documents/weather1/20170125-en.pdf, checked on 6/19/2017.

Gross, Catherine (2007): Community Perspectives of Wind Energy in Australia: The Application of a Justice and Community Fairness Framework to Increase Social Acceptance. In *Energy Policy* 35 (5), pp. 2727–2736.

Hajer, Maarten (1993): Discourse Coalitions and the Institutionalization of Practice: The Case of Acid Rain in Great Britain. In Frank Fischer, John Forester (Eds.): The Argumentative Turn in Policy Analysis and Planning. Durham, NC: Duke University Press, pp. 43–76.

Hajer, Maarten (1995): The Politics of Environmental Discourse. Ecological Modernization and the Policy Process. Oxford: Clarendon Press.

Hake, Jürgen-Friedrich; Fischer, Wolfgang; Venghaus, Sandra; Weckenbrock, Christoph (2015): The German Energiewende – history and status quo. In *Energy* 92, pp. 532–546. DOI: 10.1016/j.energy.2015.04.027.

Harris; Melissa; Beck; Marisa; Gerasimchuk; Ivetta (2015): The End of Coal: Ontario's Coal Phase-Out. International Institute for Sustainable Development. Winnipeg, Manitoba. Available online at www.iisd.org/sites/default/files/publications/end-of-coal-ontario-coal-phase-out.pdf, checked on 5/13/2016.

Hay, C. (1997): Divided by a Common Language. Political Theory and the Concept of Power. In *Politics* 17 (1), pp. 45–52. DOI: 10.1111/1467-9256.00033.

Hay, Colin (2007): Why we Hate Politics. Cambridge, Malden, MA: Polity Press.

Hill, Stephen; Knott, James (2010): Too Close for Comfort: Social Controversies Surrounding Wind Farm Noise Setback Policies in Ontario. In *Renewable Energy Law & Policy Review* (153), pp. 153–168.

IESO (2017): 50 MW Procurement Target Reached – December 1, 2017. Independent Electricity System Operator. Available online at www.ieso.ca/en/get-involved/microfit/news-bi-weekly-reports/50-mw-procurement-target-reached--december-1-2017, checked on 12/5/2017.

Kitschelt, Herbert P. (1986): Political Opportunity Structures and Political Protest: Anti-Nuclear Movements in Four Democracies. In *British Journal of Political Science* 16 (1), pp. 57–85.

Koopmans, Ruud; Olzak, Susan (2004): Discursive Opportunities and the Evolution of Right-Wing Violence in Germany. In *American Journal of Sociology* 110 (1), pp. 198–230. DOI: 10.1086/386271.

Koopmans, Ruud; Statham, Paul (1999): Ethnic and Civic Conceptions of Nationhood and the Differential Success of the Extreme Right in Germany and Italy. In Marco Giugni (Ed.): How Social Movements Matter. Minneapolis, MN: University of Minnesota Press (Social movements, protest, and contention, 10), pp. 225–252.

Landtag Brandenburg (2015): Beschlussempfehlung und Bericht des Hauptausschusses zu der Volksinitiative nach Artikel 76 der Verfassung des Landes Brandenburg „Volksinitiative für größere Mindestabstände von Windrädern sowie keine Windräder im Wald". Drucksache 6/2593.

Landwehr, Claudia; Böhm, Katharina (2014): Strategic Institutional Design: Two Case Studies of Non-Majoritarian Agencies in Health Care Priority-Setting. In *Government & Opposition*, pp. 1–29. DOI: 10.1017/gov.2014.37.

Lockwood, M.; Kuzemko, C.; Mitchell, C.; Hoggett, R. (2016): Historical Institutionalism and the Politics of Sustainable Energy Transitions. A Research Agenda. In *Environment and Planning C: Government and Policy*. DOI: 10.1177/0263774X16660561.

MacArthur, Julie. L. (2016): Challenging Public Engagement. Participation, Deliberation and Power in Renewable Energy Policy. In *Journal of Environmental Studies and Sciences* 6 (3), pp. 631–640. DOI: 10.1007/s13412-015-0328-7.

Mander, Sarah (2008): The role of discourse coalitions in planning for renewable energy: a case study of wind-energy deployment. In *Environmental Planning C* 26 (3), pp. 583–600.

MIR; MLUV (2009): Hinweise an die Regionalen Planungsgemeinschaften zur Festlegung von Eignungsgebieten „Windenergie". Windkrafterlass Brandenburg. Ministerium für Infrastruktur und Raumordnung des Landes Brandenburg (MIR); Ministerium für Ländliche Entwicklung, Umwelt und Verbraucherschutz (MLUV). Potsdam.

MUGV (2011): Beachtung naturschutzfachlicher Belange bei der Ausweisung von Windeignungsgebieten und bei der Genehmigung von Windenergieanlagen. Erlass des Ministeriums für Umwelt, Gesundheit und Verbraucherschutz vom 01. Januar 2011. Ministerium für Umwelt, Gesundheit und Verbraucherschutz.

MUNR (1994): Windkraftnutzung in Brandenburg. Ministerium für Umwelt, Naturschutz und Raumordnung des Landes Brandenburg. In *Brandenburger Umweltjournal* 4 (2), p. 14.

MUNR (1996): Windkrafterlaß für Interessenausgleich von Naturschutz und Umweltschutz. Ministerium für Umwelt, Naturschutz und Raumordnung des Landes Brandenburg. In *Brandenburger Umweltjournal* 6 (20), p. 9.

MW (2008): Energiestrategie 2020 des Landes Brandenburg. Ministerium für Wirtschaft des Landes Brandenburg. Potsdam. Available online at www.energie.brandenburg.de/media_fast/bb1.a.2755.de/Energiestrategie_2020.pdf, checked on 8/28/2012.

MWE (2012): Energiestrategie 2030 des Landes Brandenburg. Potsdam.

OCAA (2017): Ontario's Greenhouse Gas Reduction Options: A Cost Comparison. Ontario Clean Air Alliance. Available online at www.cleanairalliance.org/wp-content/uploads/2017/04/ghg-costs-APRIL20.pdf, checked on 12/11/2017.

Ontario Ministry of Energy (2013): Achieving Balance – Ontario's Long-Term Energy Plan. Queen's Printer for Ontario. Toronto.

Ontario Ministry of Energy (2017): Delivering Fairness and Choice. 2017 Long-Term Energy Plan. Queen's Printer for Ontario. Toronto.

Partzsch, Lena (2015): Kein Wandel ohne Macht – Nachhaltigkeitsforschung braucht ein mehrdimensionales Machtverständnis. In *GAIA – Ecological Perspectives for Science and Society* 24 (1), pp. 48–56. DOI: 10.14512/gaia.24.1.10.

Pierpont, Nina (2009): Wind Turbine Syndrome. A Report on a Natural Experiment. Santa Fe, N.M.: K-Selected Books.

Rucht, Dieter (1996): The Impact of National Contexts on Social Movement Structures: A Cross-Movement and Cross-National Comparison. In Doug McAdam, John D. McCarthy, Mayer N. Zald (Eds.): Comparative Perspectives on Social Movements: Political Opportunities, Mobilizing Structures, and Cultural Framings. Cambridge: Cambridge University Press, pp. 185–204.

Schmidt, Vivien A. (2008): Discursive Institutionalism: The Explanatory Power of Ideas and Discourse. In *Annual Review of Political Science* 11 (1), pp. 303–326. DOI: 10.1146/annurev.polisci.11.060606.135342.

Schmidt, Vivien A. (2010): Taking Ideas and Discourse Seriously: Explaining Change through Discursive Institutionalism as the Fourth 'New Institutionalism'. In *European Political Science Review* 2 (01), p. 1. DOI: 10.1017/S175577390999021X.

Schmidt, Vivien A. (2017): Theorizing Ideas and Discourse in Political Science. Intersubjectivity, Neo-Institutionalisms, and the Power of Ideas. In *Critical Review* 29 (2), pp. 248–263. DOI: 10.1080/08913811.2017.1366665.

Simcock, Neil (2016): Procedural Justice and the Implementation of Community Wind Energy Projects. A case study from South Yorkshire, UK. In *Land Use Policy* 59, pp. 467–477. DOI: 10.1016/j.landusepol.2016.08.034.

Snow, David A.; Benford, Robert D. (1992): Master Frames and Cycles of Protest. In Aldon D. Morris (Ed.): Frontiers in social movement theory. New Haven, CT: Yale University Press, pp. 135–155.

Songsore, Emmanuel; Buzzelli, Michael (2014): Social Responses to Wind Energy Development in Ontario: The Influence of Health Risk Perceptions and Associated Concerns. In *Energy Policy* 69, pp. 285–296. DOI: 10.1016/j.enpol.2014.01.048.

SPD Brandenburg (2015): Akzeptanz der Windenergie sichern. Beschluss des SPD-Landesvorstandes vom 7. Juli 2015. Potsdam.

SPD Brandenburg; DIE LINKE Brandenburg (2014): Sicher, Selbstbewusst und Solidarisch: Brandenburgs Aufbruch vollenden. Koalitionsvertrag zwischen SPD Brandenburg und DIE LINKE Brandenburg für die 6. Wahlperiode des Brandenburger Landtages 2014 bis 2019.

Szarka, Joseph (2004): Wind Power, Discourse Coalitions and Climate Change: Breaking the Stalemate? In *European Environment* 14 (6), pp. 317–330.

Unwilling Hosts (2015): Ontario Unwilling Hosts. Available online at http://ontario-unwilling-hosts.org/, updated on 8/13/2015, checked on 12/10/2015.

Vrablikova, Katerina (2014): How Context Matters? Mobilization, Political Opportunity Structures, and Nonelectoral Political Participation in Old and New Democracies. In *Comparative Political Studies* 47 (2), pp. 203–229. DOI: 10.1177/0010414013488538.

Walker, Chad; Baxter, Jamie (2017): Procedural Justice in Canadian Wind Energy Development. A Comparison of Community-Based and Technocratic Siting Processes. In *Energy Research & Social Science* 29, pp. 160–169. DOI: 10.1016/j.erss.2017.05.016.

Walker, Chad; Stephenson, Laura; Baxter, Jamie (2018): "His main platform is 'stop the turbines'": Political discourse, partisanship and local responses to wind energy in Canada. In *Energy Policy* 123, pp. 670–681. DOI: 10.1016/j.enpol.2018.08.046.

Wind Concerns Ontario (2015): About Us. Available online at www.windconcerns ontario.ca/about-us/, checked on 10/21/2015.

Wolsink, Maarten (2007): Planning of Renewables Schemes: Deliberative and Fair Decision-Making on Landscape Issues Instead of Reproachful Accusations of Non-Cooperation. In *Energy Policy* 35 (5), pp. 2692–2704. DOI: 10.1016/j.enpol.2006.12.002.

7 Conclusion

The changing winds of discourses on decarbonization

Many countries now put forward ambitious plans for decarbonization. Different discursive energy spaces imply different considerations of the options. The Committee on Climate Change for example, the UK's independent advisory body to the government, proposed a strategy on how to achieve a 95–96 percent greenhouse gas reduction by 2050, compared to 1990 levels (CCC 2019). The strategy focuses on reaching this target by domestic efforts and builds upon the large-scale use of carbon capture and storage (CCS) in industry and in combination with bioenergy as well as on nuclear power. The land use sector plays a key role. The scenarios of the report involve a fifth of the UK agricultural area being used for planting trees, energy crops and peatland restoration. By contrast, the German Environment Agency published a study with a similar greenhouse gas reduction target for Germany but built on entirely different assumptions (UBA 2014, 2019). Energy savings, energy efficiency gains and the use of renewable energy sources are shown to be sufficient measures to reach nearly full decarbonization by 2050. CCS, nuclear power or the cultivation of biomass crops solely to generate energy are excluded from the scenario because they are considered "not parts of a sustainable energy system" (UBA 2019, p. 42).

This shows inherently different understandings of how to reach decarbonization. In each way, decarbonization will potentially go along with land use conflicts comparable to disputes over wind turbines. The lessons of this book are therefore highly relevant for a successful decarbonization to avoid dangerous climate change. This chapter therefore first summarizes the key findings and then focuses on three relevant aspects that will need attention for the successful future implementation of decarbonization. Those revolve around the practical arrangements of the space of participation, the influence of rising populism changing the meaning context and the inhibiting role of incumbent industries. The chapter concludes with an outlook on changing discursive energy spaces in the light of accelerating climate change.

Summary of the argument

Countries are increasingly turning to renewable energy as they move their energy systems away from conventional sources such as oil, gas and nuclear. In many countries, the initiative to support renewables has been pushed by environmental social movements, exemplified by the German anti-nuclear movement and the campaign for the Green Energy Act in Ontario, Canada. Because of its relatively low cost of generation, onshore wind power has gained particular popularity as a renewable energy source. While many governments regard wind power as a convenient way to increase the share of renewable energy in their electricity system, wind turbines often meet with skepticism or even outright resistance by those who live where they are built. These different viewpoints resonate with the suggestion that long-term energy transitions "will prove to be a messy, conflictual, and highly disjointed process" (Meadowcroft 2009, p. 323). The book posed the general question of whether anti-wind movements are now on track to stop the global turn towards renewables.

Taking a comparative case study approach between two sub-national forerunner jurisdictions in installed wind power capacity, Brandenburg in Germany and Ontario in Canada, this book investigated the contributing factors that lead governments to respond differently to anti-wind movements. The study focused on the time period between 2009/2010 and 2015/2016. While the Ontario government first substantially altered and then ultimately curbed its renewable energy policy, the Brandenburg state government was reluctant in adopting the movement's demands and continued their policy approach to wind energy.

The book departed from the observation that the literature on social movements, despite its valuable contributions to understanding movement outcomes, does not provide sufficient explanations for different government responses. It has tended to focus on the separate aspects of internal movement factors (Gamson 1975), the political opportunity structure (Bloom 2014; Eisinger 1973; Kitschelt 1986) or the discursive environment in the form of media discourses (Aydemir and Vliegenthart 2017; Koopmans and Statham 1999; Motta 2015). The study drew upon this branch of literature and combined it with elements of discursive institutionalism, which is part of the "new institutionalism" branch of institutional theory. Discursive approaches to policy-making and policy contestation have been used widely in the study of environmental politics (Dryzek 1997, 2013; Feindt and Oels 2005; Hajer 1993; Leipold et al. 2019). This perspective highlights the importance of discourses in the evolution and contestation of renewable energy support policies. Yet, analyzing anti-wind disputes does not only require focusing on discourses. Especially in cross-country comparisons, the formal institutional aspects are key to identifying the possible reasons behind different government responses. This study therefore adopted a discursive-institutional (Schmidt 2008, 2010, 2012) and an argumentative discourse analysis viewpoint (Hajer 1995, 2006) to take into account both discursive and institutional aspects and explore their effect on the outcomes of anti-wind movements.

Developing the concept of "discursive energy space", this study showed empirically that differences in the discursive-institutional setting of a given jurisdiction can contribute to different government responses. The book revolved around the argument that different discursive energy spaces provide social movements with different degrees of discursive strength, influencing their ability to have their concerns taken up by the government. Discursive strength of a movement was defined as the ability to scale-up the movement's demands from localized conflicts to the sub-national policy-making level by being regarded by the government as legitimate spokespersons for a valid set of interests. The discursive energy space is conceptually made up of the *meaning context*, which includes the different discourses on energy transitions and framings of the energy sector that prevail in a jurisdiction. The discursive energy space further comprises the *formal institutional context*, which consists of the space of participation and the way in which the contested sub-national energy policy is embedded in the federal system. The space of participation encompasses local residents' participation opportunities in local wind turbine siting. These components of the discursive energy space have different power effects, which were described by drawing on simplified versions of the first (Dahl 1957), second (Bachrach and Baratz 1962) and fourth faces of power (Foucault 2000).

The comparison of the case studies suggested that Ontario's discursive energy space provided more discursive strength to the anti-wind movement than Brandenburg's. The prevailing meaning context in Ontario allowed the anti-wind movement to successfully frame wind turbines in terms of a severe health threat, while the Brandenburg movement could not make health a central topic of their campaign even though it figured prominently among their concerns. This can be explained by the differences between Brandenburg and Ontario's meaning contexts. In contrast to Brandenburg, the Ontario energy sector had been strongly linked to health questions before. This is exemplified by the successful campaign for Ontario's 2014 coal phase-out, which was mainly informed by the discourse on air pollution (Harris et al. 2015). Hence, the characteristics of the meaning context made it easier for the Ontario movement to mobilize support on the health frame.

In Brandenburg, the opposite was the case. Brandenburg's meaning context is heavily shaped by the German discourse of *Energiewende*, which connects to the Brandenburg discourse of *Energieland*. The *Energiewende* discourse is widely accepted in German society and implies the need to build an energy system based on renewables in order to phase-out nuclear in the short and fossil fuels in the medium-term. The *Energieland* discourse is mainly used by the Brandenburg government to highlight Brandenburg's historic importance as a major energy producer during GDR times and the continuing economic importance of its large lignite industry. In light of the need to phase out lignite one day, the development of renewable energy should secure ongoing energy exports and therefore also contribute to the *Energiewende* (MWE 2012; SPD Brandenburg and DIE LINKE Brandenburg 2014). Both discourses of *Energiewende* and *Energieland* weakened the anti-wind movement in Brandenburg by providing discursive references to the state government to reject the movement's demands. The demands included a separation distance between wind turbines and residential buildings of ten times the height of a wind turbine ("10h") and a ban on building wind turbines in forests. As the research in Brandenburg revealed, addressing these demands was regarded as compromising Brandenburg's ambitions to contribute to the *Energiewende* and remain an *Energieland*. The importance of the *Energiewende* discourse in Brandenburg's meaning context also contributed to the observation that the major opposition party, the CDU, was reluctant to fully support the anti-wind movement and become their strategic ally as in Ontario. This was because some members of the anti-wind movement were outspoken climate skeptics, which is not deemed a legitimate position within the meaning context of the *Energiewende*. In Ontario, by contrast, the major opposition party fully supported the anti-wind movement because no such restriction existed and wind turbines were not part of an overall discourse on the need for energy transition. The meaning context in Ontario thus added to the movement's discursive strength; in contrast, the *Energiewende* and *Energieland* discourse of Brandenburg's meaning context weakened the movement's discursive strength.

Ontario's anti-wind movement did not only gain discursive strength from the meaning context. The formal institutional context, as the second component of the discursive energy space, also contributed to the stronger force of the Ontario movement. Ontario has sovereignty over its energy sector and the studied time period fell under the federal government of Prime Minister Stephen Harper, which was characterized by weak federal leadership in the realms of climate change and environmental protection. There were thus fewer opportunities for the Ontario government to explain a possible inaction towards addressing the anti-wind movement's concerns by the need to contribute to federal level programs or targets. Nor was it possible for the Ontario government to refer to the federal level to deal with some of the movement's demands in place of itself. Yet, both ways to refer to the federal level were possible in Brandenburg. In Brandenburg, the government was reluctant to adopt the movement's demands because they were regarded as compromising the responsibility of Brandenburg to contribute to the renewable energy deployment targets at the German federal level. Furthermore, the Brandenburg government called upon the federal level to deal with rising electricity costs, which the government identified as being one of the major causes for public resistance.

The second component of the discursive energy space's formal institutional system, the space of participation, worked in a similar way. It provided more strength to the anti-wind movement in Ontario than to the Brandenburg movement. This is because the Ontario government decided to replace the previously existing decision-making system of wind turbine siting with a streamlined approach upon the implementation of its major policy for renewable energy development, the Green Energy Act. The change consisted of superseding the municipal level and restricting the reasons for appealing a project to only two grounds: proving either "(a) serious harm to human health; or (b) serious and irreversible harm to plant life, animal life or the natural environment" (Environmental Review Tribunal 2015). The then Prime Minister Dalton McGuinty introduced the changes with the remark: "NIMBYism will no longer prevail" (Ferguson and Ferenc 2009). This new streamlined approach sparked a discourse on "wind turbines are undemocratic" and also made many township councils across Ontario join the anti-wind movement's cause and declare their municipality as "unwilling host to wind turbines" (Unwilling Hosts 2015). The contrary applies to Brandenburg where the government has implemented its renewable energy policies, the Energy Strategy 2020 and 2030 (MW 2008; MWE 2012), by endorsing its existing system of regional planning. This system offers more options for local input than Ontario's, but the process is at the same time heavily restricted by state guidance, national and European legislation as well as corresponding court rules. Despite the limited actual possibilities for local participation, the government successfully framed the regional planning approach as a local and inclusive process and referred to it, as opposed to the sub-national governmental level, as being best suited to address some of the movement's demands. The space of participation therefore did not prompt a similarly strong reaction by municipal actors as in Ontario, although their possibilities to influence decisions over wind turbines in their territories were also severely restricted.

While the two components of the discursive energy space thus contributed to a stronger discursive strength of the Ontario anti-wind movement as compared to the Brandenburg movement, the study also suggests that the meaning context and the formal institutional context interrelate. First, the *Energiewende* discourse in Brandenburg was also institutionalized in the overarching federal legislation. As Ontario's renewable energy policy was independent from federal energy policy and there was no comparably strong discourse on energy transition, which all major parties agreed upon as the *Energiewende* and *Energieland* discourse in Brandenburg, the dispute over wind turbines became a major issue of party politics and competition. In the run-up to the 2011 provincial election, the Progressive Conservatives ran on a platform promising to place a moratorium on wind turbines until health studies were completed, re-establish local planning authority and end the feed-in tariff for renewable energy altogether. The leader of the opposition party regularly met with the president of the anti-wind umbrella organization Wind Concerns Ontario. The Progressive Conservatives therefore became a strategic ally for the anti-wind movement. This put more pressure on the government by providing the anti-wind movement with more leverage to scale-up their concerns from localized debates to provincial politics. The study further suggested that the space of participation, as a central component of the formal institutional system, also sparked discourses that became highly relevant in the overall conflict. Ontario's anti-wind movement based their campaign on the "rights" issue, referring to the lifting of decision-making authority over wind turbines from the municipal to the provincial level in the context of the implementation of the Green Energy Act. This suggests that discourses stemming from formal institutions such as the space of participation can become highly relevant in disputes over a specific policy. Looking at both the meaning context and the discursive effects of the institutional context represents the major idea behind the concept of discursive energy space. The study suggested that the power effects of the discursive energy space affects movement discursive strength.

The first face of power was relevant for the space of participation that directly excluded particular concerns from being considered. The second form of power was relevant for all components of the discursive energy space. The Brandenburg government that was confronted with anti-wind opposition referred to particular discourses (meaning context), to the federal level (role of federal energy policy), or to the local space of participation (space of participation) to reject addressing the movement's demands. This reference of governments to another level of decision-making has also been called "institutional" or "governmental depoliticization" or "strategic institutional design" (Flinders and Buller 2006; Flinders and Wood 2014; Hay 2007; Landwehr and Böhm 2014). The referral to another level is effectively used to "shape the context" of decision-making (Hay 1997) and thereby "limit[s] the scope of the political process" (Bachrach and Baratz 1962, p. 948). The fourth face of power considers Foucault's notion of discursive power (Foucault 2000) and was relevant with regard to the meaning context. The (non)existence of powerful discourses weakened or reinforced the validity of the movements' concerns and thereby affected the way in which their demands were regarded as legitimate by the government.

The new climate movement is now changing the meaning context of the discursive energy space in Germany and Canada, strengthening positive discourses on decarbonization. For the time being, it is not clear whether this influence will be strong enough to shift politics in Ontario back towards supporting the development of renewable energy. In the same vein, it is still unclear how the new public salience of climate issues will prompt Germany to reach its renewable energy targets and decarbonize its economy. Reaching net-zero emissions has become a major policy goal in many jurisdictions, and it is hard to predict which particular measures will be most promising. As many options for decarbonization require a change in land use, it is clear, however, that land-use conflicts will continue to pose an important challenge.

Towards a successful decarbonization

Electricity generation has always been highly conflictual. Under the fossil-nuclear system, conflicts over land use were often far removed from electricity consumption and mostly concentrated on few and centralized places. Fossil and nuclear resource extraction often go along with environmental degradation and human rights violations in countries located far away from where energy is eventually generated and consumed. Uranium mining in Western Africa or Australia or coal mining in Colombia are such examples. The construction of oil pipelines on Indigenous land in Canada or the resettlement of villages in Germany to install new open cast lignite mines are also a consequence of the current fossil-nuclear dominance. As energy systems are now increasingly shifted towards renewable energy and many other decarbonization processes require vast tracts of land, this "landscape of conflict" is changing. Renewable energy facilities can be placed principally almost in each corner of a country. Instead of few big plants supplying a large share of power, numerous small facilities provide smaller percentages of the total power need. This makes energy generation more visible and moves it closer to consumption. This leads to conflicts with existing uses of space. Furthermore, it is not only renewable energy facilities that are discussed for realizing a decarbonized world. Pylons, CCS or the generation of land-intensive negative emissions to compensate for the remaining greenhouse gas emissions are widely discussed options for realizing decarbonization, next to a change in consumption patterns, energy savings and efficiency. In most of the technology-based decarbonization options, the rural areas bear a large share of the burden which contributes to conflict. The success of decarbonization will, to a certain degree, also depend on the solutions that are found for bridging this urban/rural divide.

Several lessons can be drawn from this book for the fate of renewable energies and other technology-based decarbonization projects. Their success will depend on a number of aspects that relate to institutional factors, most prominently the way in which such systems are implemented. Their success also depends on the extent to which societal discourses are in favor of decarbonization or how they will be changed by emerging populist forces and incumbent industries that have no interest in changing their business model. These factors will now be shortly addressed before moving to the topic of how climate change may also alter prevailing discourses in the near future.

Getting the rules right: a meaningful participation to decarbonization projects

This book has shown that institutional factors crucially shape how conflicts over renewable energy schemes unfold. While the Brandenburg and Ontario governments were successful in launching renewable energy support policies making them forerunners in wind turbine development, they failed to adequately shape their siting procedures and models of ownership in a manner inclusive of the local population. Both Brandenburg and Ontario residents experienced wind turbine development as large-scale installations with little or no community participation. This restricted space of participation can be identified as the major reason behind high levels of protest. In the literature on conflicts over wind turbines, it has indeed become a truism that the more the process of wind turbine siting is regarded as unjust by the local population, the more resistance there will be. In contrast, meaningful community participation carried out in the right way has been consistently linked to higher levels of ownership and acceptance (Goedkoop and Devine-Wright 2016; Gross 2007; MacArthur 2016; Wolsink 2007).

What lesson can be drawn for a successful implementation of renewable energy and other decarbonization projects in general? Regarding decentralized renewable energy projects, priority should be given to community projects that incorporate the idea of "energy democracy". This includes popular sovereignty, participatory governance and civic ownership (Szulecki 2017). A related concept is "energy justice", which includes distributive and procedural justice, but also justice with regard to global externalities and the recognition of vulnerable groups (Sovacool et al. 2019). Support programs and consultancy work can help with the set-up of projects that embrace these principles. Experiences from other cases and countries regarding implementation barriers and solutions to possible conflicts should be included in the design of such projects. Furthermore, adjacent policy sectors should be streamlined in order not to inhibit the meaningful participation of communities. In Germany for example, the financial participation of host municipalities to wind turbines is often made difficult by sectoral laws pertaining to the officially allowed municipal revenues. Additionally, a system perspective should be adopted that integrates the heating and transportation sector as well. This will increase credibility of renewable energy policies. A system perspective furthermore does not stop at the urban boundary. Urban areas should be encouraged to contribute to energy saving, energy efficiency and renewable energy generation in order not to place the burden exclusively on rural areas. For instance, rooftop photovoltaic is still an untapped source of power in most cities. While these recommendations represent institutional options to shape the turn towards renewable energy in an equitable way, the fate of renewable energy programs will ultimately also depend on whether favorable discourses about them prevail in a society.

Taking the wind out of populist sails

As much as the new climate movement is shifting the public discourse towards more urgent climate action, a counterforce has also become more important recently. Decarbonization efforts are increasingly being challenged by the rise of populist forces (Lockwood 2018). Regarding the two sub-national case study jurisdictions of Brandenburg and Ontario, recent elections have resulted in a strong momentum for right-wing populist forces that are questioning or stopping the development of renewable energy. In Brandenburg, the right-wing populist AfD gained an historic 23.5 percent of the vote in the 2019 state election, occupying second place after the incumbent social-democrats. Ontario's Premier Doug Ford of Ontario's Progressive Conservatives openly dismissed environmental legislation and renewable energy programs since he gained office in 2018, engaging in populist rhetoric (Lachapelle and Kiss 2019). The rise of populists is not restricted to the case study jurisdictions – the United States, Brazil, Turkey, Russia and Poland are just a few examples of other jurisdictions where environmental and climate legislation has either been crucially slashed or never gained the necessary attention due to populist politics.

Populism influences decarbonization projects in many ways. Right-wing populists offer oversimplified solutions to complex problems. They take advantage of people feeling overburdened by the multiple existing global challenges such as climate change, digitalization, migration, inequality, global disease or ecological degradation, resulting in changes in their local world and habits. Populists choose the easy answer and deny the cause of the problem, including anthropogenic climate change and instead putting forward *fossil discourses*. They thereby become natural allies of those who question the need for decarbonization, including renewable energy. The current Premier of Ontario is an example of how renewable energy projects and economic instruments for reducing CO_2 emissions are cancelled if a government institutionalizes *fossil discourses*, even if not directly denying anthropogenic climate change.

Right-wing populists attack some of the basic principles of democracy including reliance on scientific facts, making compromises and speaking the truth. This puts democratic systems under pressure. The fact that populist forces are often represented by official political parties eligible for election conveys the general impression that what they say and do, even if it denies scientific findings, is an acceptable perspective. As it shifts the boundaries of what can be said and done, this changes the meaning context and thereby has far-reaching consequences on how societal problems are addressed and solutions found. The 2019 electoral program of the AfD in Brandenburg for example explained climate change by bold and unproven theories, demanding an end to climate change policies and the support of renewable energy. Calling themselves the "alternative" suggests that drawing on false theories just represents another possible option in a pluralist political system. This runs the risk of altering the meaning context in a way that scientific findings become negotiable and dependent on one's own opinion.

As not less than the very functioning of democracy is at stake, it is important that democratic forces in a society find a counter-strategy in order to take the wind out of populist sails. This must include refusal to give in to their argumentation as well as a comprehensive social media strategy, as this is the channel where populist forces have become very active and successful in getting their message across. Taking the wind out of populists' sails may be easier said than done, and since many democratic systems are faced with this challenge, research should focus more on this topic.

Defying incumbent industries

Closely related with the rise of populist forces is the role of incumbent industries. Recent populist politics most often leans towards the fossil fuel sector and the accompanying *fossil discourse*, openly questioning anthropogenic climate change and denying the need to decarbonize energy systems (Lockwood 2018). There are also many instances of conservative political parties protective of their jurisdiction's fossil fuel sector. Apart from this protective attitude towards fossil-fuel based energy systems, the fossil and nuclear sector also play a role in energy system change by themselves (Edberg and Tarasova 2016; Kooij et al. 2018; Kungl 2015; Wodrig 2018). The case studies suggested that the fossil and nuclear sectors were both drivers of energy system change, but also inhibitors of such change. In Brandenburg and Ontario, actual or anticipated changes in the fossil and nuclear sectors served as major motivations to invoke the *ecological modernization* discourse. The governments in Brandenburg and Ontario established their ambitious renewable energy programs as win-win situations between economic, ecological and partly community-focused benefits. The *fossil discourse* has prevailed for a long time in Brandenburg because its economy strongly benefitted from lignite-fired energy generation. It could be argued that only because Brandenburg had such strong ties to lignite mining was it open to fostering renewable energy to such a great extent. In Ontario, the coal phase-out also served as a major source of legitimation for introducing renewable energy targets. Yet, the fossil and nuclear sectors are not easily replaced by a new industry. In Brandenburg, the introduction of renewables did not directly lead to a coal phase-out and in Ontario, the strong role of nuclear power was in the end one of the decisive factors for terminating the renewable energy program. Meeting power supply with the refurbishment of nuclear power stations was less contentious than developing more wind energy. While the study thus suggested some links between the incumbent sector and the turn to renewables, further avenues for research could focus more directly on the power effects of the incumbent sector to actively shape a jurisdiction's energy policy. This touches upon existing studies on the role of the incumbent industries for energy and this branch of literature could be strengthened by further research. Given the fact that most of current climate research funding is spent in the natural science (Overland and Sovacool 2020), more research should generally be directed towards climate research in the area of social science, including the role of interest groups such as incumbents, the rise of populism and the effect of shifting discourses on decarbonization.

Changing climate, changing discourses: the way forward

Anti-wind movements have developed considerably since they first emerged. Taking the shape of isolated protests in the first years of wind turbine development, they soon organized into local and regional protest groups and formed umbrella organizations. They take advantage of social media to network, exchange information and spread their message. Many instances of anti-wind movements now reach a high degree of professionalization, visibility and impact. Despite the fact that they often represent a small, but vocal share of the population, some instances of anti-wind movements have become powerful enough to prompt a change in energy policies. At the time the research for this book was undertaken, there was no clear evidence that only a couple of years later, a new, rather urban and young climate movement would emerge, gaining considerable public attention and exerting pressure on governments to act on climate change. This new climate movement demands rapid action to reduce emissions, decarbonize energy systems and increase the share of renewable energy. It contributes to a changing discourse on the importance of climate action and therefore to a changing meaning context of disputes over wind turbines. Those who are in favor of wind power may benefit from this increased salience of climate issues, while those arguing against wind power may have a harder time finding an audience to listen to them. For the future, it remains to be seen whether the goals of the new climate movement will prevail over the concerns of the social movements against wind power, leading to continued or renewed support for renewable energy.

The new climate movement is likely to gain more momentum in the future. As the impacts of climate change will become more visible also for industrialized Western countries including North America and Europe, so will the need for mitigation become more prominent and the options for decarbonization more intensely debated. The current scenario based on policies presently in place around the world is a world that is on track to become three degrees warmer (Climate Action Tracker 2019). To name just a few impacts, increasing weather extremes, both in intensity and frequency, lead to heavy rainfalls, flooding and heat waves. The spread of tropical disease challenges health care systems. Faced with these disruptive developments, people may realize that the costs for mitigation are lower than for adaptation and emergency assistance. The same applies for the health argument: When health-related climate impacts become more apparent, arguments against decarbonization will undergo a new evaluation. Ultimately, these new insights may lead to serious decarbonization efforts, including behavioral changes such as a change in consumption patterns and energy savings, but also technology-based measures such as the increase of renewable energy and, to a certain extent, also CCS and negative emissions technologies.

Despite this case for a new assessment of decarbonization impacts, successful decarbonization efforts depend on political forces that take them seriously. There might also be the scenario that in the absence of a positive narrative on a decarbonized future, right-wing populism offering easy explanations and solutions gains strength. In this scenario, populist forces are in government or, as political opposition parties, raise the political costs for incumbent governments to engage in ambitious and effective decarbonization. Consequently, governments opt to rather do nothing or insufficiently address deep decarbonization. This may contribute to a downward spiral of the Paris Agreement (Sachs 2019) and its possible ultimate failure to meet the temperature targets. Industrialized countries will then have to spend an important share of their financial resources to adapt to climate change. Millions of people in the global South will face poor living conditions and are likely to migrate. As costs for adaptation to climate change will escalate and the number of directly affected people will increase, this political "no-strategy" comes at a high social cost and may even result in civil unrest.

In reality, the future is likely to be a mix of these two scenarios. As strong as the counterforces to effective climate change mitigation may seem at times, humanity as a species is able to learn and change direction. Maybe it is only in the face of the most severe impacts that discourses will change and real climate mitigation will take place.

References

Aydemir, Nermin; Vliegenthart, Rens (2017): Public Discourse on Minorities. How Discursive Opportunities Shape Representative Patterns in the Netherlands and the UK. In *Nationalities Papers* 4 (2), pp. 1–15. DOI: 10.1080/00905992.2017.1342077.

Bachrach, Peter; Baratz, Morton S. (1962): Two Faces of Power. In *American Political Science Review* 56 (04), pp. 947–952. DOI: 10.2307/1952796.

Bloom, Jack M. (2014): Political Opportunity Structure, Contentious Social Movements, and State-Based Organizations. The Fight against Solidarity inside the Polish United Workers Party. In *Social Science History* 38 (3–4), pp. 359–388. DOI: 10.1017/ssh.2015.29.

CCC (2019): Net Zero – The UK's Contribution to Stopping Global Warming – Committee on Climate Change. Committee on Climate Change. Available online at www.theccc.org.uk/publication/net-zero-the-uks-contribution-to-stopping-global-warming/#more-information, checked on 1/14/2020.

Climate Action Tracker (2019): Global update: Governments Still Showing Little Sign of Acting on Climate Crisis | Climate Action Tracker. Available online at https://climate-actiontracker.org/press/global-update-governments-showing-little-sign-of-acting-on-climate-crisis/, checked on 1/2/2020.

Dahl, Robert (1957): The Concept of Power. In *Behavioral Science* 2 (3), pp. 201–215.

Deutscher Bundestag (2018): Evaluierungsbericht der Bundesregierung über die Anwendung des Kohlendioxid-Speicherungsgesetzes sowie die Erfahrungen zur CCS-Technologie. Drucksache 19/6891.

Dryzek, John S. (1997): The Politics of the Earth. Environmental Discourses. Oxford, New York: Oxford University Press.

Dryzek, John S. (2013): The Politics of the Earth. Environmental Discourses. 3. ed. Oxford: Oxford University Press.

Edberg, Karin; Tarasova, Ekaterina (2016): Phasing out or phasing in. Framing the role of nuclear power in the Swedish energy transition. In *Energy Research & Social Science* 13, pp. 170–179. DOI: 10.1016/j.erss.2015.12.008.

Eisinger, Peter K. (1973): The Conditions of Protest Behavior in American Cities. In *The American Political Science Review* 67 (1), pp. 11–28. Available online at www.jstor.org/stable/pdf/1958525.pdf.

Environmental Review Tribunal (2015): Case No. 14–096 Mothers Against Wind Turbines Inc. v. Ontario (Environment and Climate Change). Proceeding commenced under section 142.1(2) of the Environmental Protection Act, R.S.O. 1990, c.E.19, as amended. Environmental Review Tribunal. Available online at www.ert.gov.on.ca/english/hearings/index.htm, checked on 6/20/2015.

Feindt, Peter H.; Oels, Angela (2005): Does Discourse Matter? Discourse Analysis in Environmental Policy Making. In *Journal of Environmental Policy & Planning* 7 (3), pp. 161–173. DOI: 10.1080/15239080500339638.

Ferguson, Rob; Ferenc, Leslie (2009): McGuinty Vows to Stop Wind-Farm NIMBYs. In *Toronto Star*, 2/11/2009. Available online at www.thestar.com/news/ontario/2009/02/11/mcguinty_vows_to_stop_windfarm_nimbys.html, checked on 11/2/2015.

Flinders, Matthew; Buller, Jim (2006): Depoliticisation. Principles, Tactics and Tools. In *British Politics* 1 (3), pp. 293–318. DOI: 10.1057/palgrave.bp.4200016.

Flinders, Matthew; Wood, Matt (2014): Depoliticisation, Governance and the State. In *Policy & Politics* 42 (2), pp. 135–149. DOI: 10.1332/030557312X655873.

Foucault, Michel (2000): Truth and Power. In Michel Foucault, Paul Rabinow, James D. Faubion (Eds.): The Essential Works of Michel Foucault, 1954–1984. New York: New Press (The essential works of Michel Foucault, 1954–1984, v. 1).

Gamson, William A. (1975): The Strategy of Social Protest. Homewood Ill.: Dorsey Press (The Dorsey series in sociology).

Goedkoop, Fleur; Devine-Wright, Patrick (2016): Partnership or Placation? The Role of Trust and Justice in the Shared Ownership of Renewable Energy Projects. In *Energy Research & Social Science* 17, pp. 135–146. DOI: 10.1016/j.erss.2016.04.021.

Gross, Catherine (2007): Community Perspectives of Wind Energy in Australia. The Application of a Justice and Community Fairness Framework to Increase Social Acceptance. In *Energy Policy* 35 (5), pp. 2727–2736.

Hajer, Maarten (1993): Discourse Coalitions and the Institutionalization of Practice. The Case of Acid Rain in Great Britain. In Frank Fischer, John Forester (Eds.): The Argumentative Turn in Policy Analysis and Planning. Durham, N.C: Duke University Press, pp. 43–76.

Hajer, Maarten (1995): The Politics of Environmental Discourse. Ecological Modernization and the Policy Process. Oxford: Clarendon Press.

Hajer, Maarten (2006): Doing Discourse Analysis. Coalitions, Practices, Meaning. In Brink, Margo van den (Ed.): Words Matter in Policy and Planning. Discourse Theory and Method in the Social Sciences. Utrecht: Koninklijk Nederlands Aardrijkskundig Genootschap (Netherlands geographical studies, 344), pp. 65–76. Available online at www.maartenhajer. nl/images/stories/20080204_MH_wordsmatter_ch4.pdf, checked on 12/1/2014.

Harris; Melissa; Beck; Marisa; Gerasimchuk; Ivetta (2015): The End of Coal: Ontario's Coal Phase-Out. International Institute for Sustainable Development. Winnipeg, Manitoba. Available online at www.iisd.org/sites/default/files/publications/end-of-coal-ontario-coal-phase-out.pdf, checked on 5/13/2016.

Hay, C. (1997): Divided by a Common Language. Political Theory and the Concept of Power. In *Politics* 17 (1), pp. 45–52. DOI: 10.1111/1467-9256.00033.

Hay, Colin (2007): Why we Hate Politics. Cambridge, Malden, MA: Polity Press.

Kitschelt, Herbert P. (1986): Political Opportunity Structures and Political Protest: Anti-Nuclear Movements in Four Democracies. In *British Journal of Political Science* 16 (1), pp. 57–85.

Kooij, Henk-Jan; Oteman, Marieke; Veenman, Sietske; Sperling, Karl; Magnusson, Dick; Palm, Jenny; Hvelplund, Frede (2018): Between grassroots and treetops. Community power and institutional dependence in the renewable energy sector in Denmark, Sweden and the Netherlands. In *Energy Research & Social Science* 37, pp. 52–64. DOI: 10.1016/j.erss.2017.09.019.

Koopmans, Ruud; Statham, Paul (1999): Ethnic and Civic Conceptions of Nationhood and the Differential Success of the Extreme Right in Germany and Italy. In Marco Giugni (Ed.): How Social Movements Matter. Minneapolis Minn.: Univ. of Minnesota Press (Social movements, protest, and contention, 10), 225–252.

Kungl, Gregor (2015): Stewards or Sticklers for Change? Incumbent Energy Providers and the Politics of the German Energy Transition. In *Energy Research & Social Science* 8, pp. 13–23. DOI: 10.1016/j.erss.2015.04.009.

Lachapelle, Erick; Kiss, Simon (2019): Opposition to Carbon Pricing and Right-wing Populism. Ontario's 2018 general election. In *Environmental Politics* 28 (5), pp. 970–976. DOI: 10.1080/09644016.2019.1608659.

Landwehr, Claudia; Böhm, Katharina (2014): Strategic Institutional Design. Two Case Studies of Non-Majoritarian Agencies in Health Care Priority-Setting. In *Government & Opposition*, pp. 1–29. DOI: 10.1017/gov.2014.37.

Leipold, Sina; Feindt, Peter H.; Winkel, Georg; Keller, Reiner (2019): Discourse Analysis of Environmental Policy Revisited. Traditions, Trends, Perspectives. In *Journal of Environmental Policy & Planning* 21 (5), pp. 445–463. DOI: 10.1080/1523908X. 2019.1660462.

Lockwood, Matthew (2018): Right-wing Populism and the Climate Change Agenda. Exploring the linkages. In *Environmental Politics* 27 (4), pp. 712–732. DOI: 10.1080/ 09644016.2018.1458411.

MacArthur, Julie. L. (2016): Challenging Public Engagement. Participation, Deliberation and Power in Renewable Energy Policy. In *Journal of Environmental Studies Science* 6 (3), pp. 631–640. DOI: 10.1007/s13412-015-0328-7.

Meadowcroft, James (2009): What About the Politics? Sustainable Development, Transition Management, and Long Term Energy Transitions. In *Policy Science* 42 (4), pp. 323–340. DOI: 10.1007/s11077-009-9097-z.

Motta, Renata (2015): Transnational Discursive Opportunities and Social Movement Risk Frames Opposing GMOs. In *Social Movement Studies* 14 (5), pp. 576–595. DOI: 10.1080/14742837.2014.947253.

MW (2008): Energiestrategie 2020 des Landes Brandenburg. Ministerium für Wirtschaft des Landes Brandenburg. Potsdam. Available online at www.energie.brandenburg.de/media_fast/bb1.a.2755.de/Energiestrategie_2020.pdf, checked on 8/28/2012.

MWE (2012): Energiestrategie 2030 des Landes Brandenburg. Potsdam.

Overland, Indra; Sovacool, Benjamin K. (2020): The Misallocation of Climate Research Funding. In *Energy Research & Social Science* 62, p. 101349. DOI: 10.1016/j.erss.2019.101349.

Sachs, Noah (2019): The Paris Agreement in the 2020s: Breakdown or Breakup? In *Ecology Law Quarterly* 46 (1).

Schmidt, Vivien A. (2008): Discursive Institutionalism: The Explanatory Power of Ideas and Discourse. In *Annual Review Political Science* 11 (1), pp. 303–326. DOI: 10.1146/annurev.polisci.11.060606.135342.

Schmidt, Vivien A. (2010): Taking Ideas and Discourse Seriously: Explaining Change through Discursive Institutionalism as the Fourth 'New Institutionalism'. In *European Political Science Review* 2 (01), p. 1. DOI: 10.1017/S175577390999021X.

Schmidt, Vivien A. (2012): Discursive Institutionalism: Scope, Dynamics, and Philosophical Underpinnings. In F. Fischer, H. Gottweis (Eds.): The Argumentative Turn Revisited: Public Policy as Communicative Practice. Durham, NC, London: Duke University Press, pp. 85–113.

Sovacool, Benjamin K.; Martiskainen, Mari; Hook, Andrew; Baker, Lucy (2019): Decarbonization and its discontents: a critical energy justice perspective on four low-carbon transitions. In *Climatic Change* 155 (4), pp. 581–619. DOI: 10.1007/s10584-019-02521-7.

SPD Brandenburg; DIE LINKE Brandenburg (2014): Sicher, Selbstbewusst und Solidarisch: Brandenburgs Aufbruch vollenden. Koalitionsvertrag zwischen SPD Brandenburg und DIE LINKE Brandenburg für die 6. Wahlperiode des Brandenburger Landtages 2014 bis 2019.

Szulecki, Kacper (2017): Conceptualizing energy democracy. In *Environmental politics* 27 (1), pp. 21–41. DOI: 10.1080/09644016.2017.1387294.

UBA (2014): Germany in 2050 – A Greenhouse Gas-neutral Country. German Environment Agency (Umweltbundesamt). Available online at www.umweltbundesamt.de/sites/default/files/medien/378/publikationen/07_2014_climate_change_en.pdf.

UBA (2019): A Resource Efficient Pathway Towards a Greenhouse Gas Neutral Germany. German Environment Agency (Umweltbundesamt). Available online at www.umweltbundesamt.de/publikationen/a-resource-efficient-pathway-towards-a-greenhouse.

Unwilling Hosts (2015): Ontario Unwilling Hosts. Available online at http://ontario-unwilling-hosts.org/, updated on 8/13/2015, checked on 12/10/2015.

Wodrig, Stefanie (2018): New Subjects in the Politics of Energy Transition? Reactivating the Northern German Oil and Gas Infrastructure. In *Environmental Politics* 27 (1), pp. 69–88. DOI: 10.1080/09644016.2017.1384469.

Wolsink, Maarten (2007): Planning of Renewables Schemes. Deliberative and Fair Decision-Making on Landscape Issues Instead of Reproachful Accusations of Non-Cooperation. In *Energy Policy* 35 (5), pp. 2692–2704.

Index

Page numbers in **bold** denote tables, those in *italics* denote figures.

For Product Safety Concerns and Information please contact our EU
representative GPSR@taylorandfrancis.com
Taylor & Francis Verlag GmbH, Kaufingerstraße 24, 80331 München, Germany

www.ingramcontent.com/pod-product-compliance
Lightning Source LLC
Chambersburg PA
CBHW060553220326
41598CB00024B/3088